Lecture Notes in Mathematics

Edited by J.-M. Morel, F. Takens and B. Teissier

Editorial Policy
for the publication of monographs

1. Lecture Notes aim to report new developments in all areas of mathematics and their applications - quickly, informally and at a high level. Mathematical texts analysing new developments in modelling and numerical simulation are welcome.

   Monograph manuscripts should be reasonably self-contained and rounded off. Thus they may, and often will, present not only results of the author but also related work by other people. They may be based on specialised lecture courses. Furthermore, the manuscripts should provide sufficient motivation, examples and applications. This clearly distinguishes Lecture Notes from journal articles or technical reports which normally are very concise. Articles intended for a journal but too long to be accepted by most journals, usually do not have this „lecture notes" character. For similar reasons it is unusual for doctoral theses to be accepted for the Lecture Notes series, though habilitation theses may be appropriate.

2. Manuscripts should be submitted (preferably in duplicate) either to Springer's mathematics editorial in Heidelberg, or to one of the series editors (with a copy to Springer). In general, manuscripts will be sent out to 2 external referees for evaluation. If a decision cannot yet be reached on the basis of the first 2 reports, further referees may be contacted: The author will be informed of this. A final decision to publish can be made only on the basis of the complete manuscript, however a refereeing process leading to a preliminary decision can be based on a pre-final or incomplete manuscript. The strict minimum amount of material that will be considered should include a detailed outline describing the planned contents of each chapter, a bibliography and several sample chapters.

   Authors should be aware that incomplete or insufficiently close to final manuscripts almost always result in longer refereeing times and nevertheless unclear referees' recommendations, making further refereeing of a final draft necessary.

   Authors should also be aware that parallel submission of their manuscript to another publisher while under consideration for LNM will in general lead to immediate rejection.

3. Manuscripts should in general be submitted in English. Final manuscripts should contain at least 100 pages of mathematical text and should always include

   - a table of contents;
   - an informative introduction, with adequate motivation and perhaps some historical remarks: it should be accessible to a reader not intimately familiar with the topic treated;
   - a subject index: as a rule this is genuinely helpful for the reader.

   For evaluation purposes, manuscripts may be submitted in print or electronic form (print form is still preferred by most referees), in the latter case preferably as pdf- or zipped ps-files. Lecture Notes volumes are, as a rule, printed digitally from the authors' files. To ensure best results, authors are asked to use the LaTeX2e style files available from Springer's web-server at:

   ftp://ftp.springer.de/pub/tex/latex/mathegl/mono/ (for monographs) and

   ftp://ftp.springer.de/pub/tex/latex/mathegl/mult/ (for summer schools/tutorials).

   Additional technical instructions, if necessary, are available on request from lnm@springer-sbm.com.

Continued on inside back-cover

Lecture Notes in Mathematics 1904

Peter Giesl

Construction of Global Lyapunov Functions Using Radial Basis Functions

Springer

Author

Peter Giesl
Centre for Mathematical Sciences
University of Technology München
Boltzmannstr. 3
85747 Garching bei München
Germany
e-mail: giesl@ma.tum.de

Library of Congress Control Number: 2007922353

Mathematics Subject Classification (2000): 37B25, 41A05, 41A30, 34D05

ISSN print edition: 0075-8434
ISSN electronic edition: 1617-9692

ISBN 978-3-540-69907-1 Springer Berlin Heidelberg New York
DOI 10.1007/978-3-540-69909-5

Springer is a part of Springer Science+Business Media
springer.com

Typesetting by the author using a Springer LaTeX macro package
Cover design: WMXDesign GmbH, Heidelberg

Printed on acid-free paper SPIN: 11979265 VA41/3100/SPi 5 4 3 2 1 0

Preface

This book combines two mathematical branches: dynamical systems and radial basis functions. It is mainly written for mathematicians with experience in at least one of these two areas. For dynamical systems we provide a method to construct a Lyapunov function and to determine the basin of attraction of an equilibrium. For radial basis functions we give an important application for the approximation of solutions of linear partial differential equations. The book includes a summary of the basic facts of dynamical systems and radial basis functions which are needed in this book. It is, however, no introduction textbook of either area; the reader is encouraged to follow the references for a deeper study of the area.

The study of differential equations is motivated from numerous applications in physics, chemistry, economics, biology, etc. We focus on autonomous differential equations $\dot{x} = f(x)$, $x \in \mathbb{R}^n$ which define a dynamical system. The simplest solutions $x(t)$ of such an equation are equilibria, i.e. solutions $x(t) = x_0$ which remain constant. An important and non-trivial task is the determination of their basin of attraction.

The determination of the basin of attraction is achieved through sublevel sets of a Lyapunov function, i.e. a function with negative orbital derivative. The orbital derivative $V'(x)$ of a function $V(x)$ is the derivative along solutions of the differential equation.

In this book we present a method to construct Lyapunov functions for an equilibrium. We start from a theorem which ensures the existence of a Lyapunov function T which satisfies the equation $T'(x) = -\bar{c}$, where $\bar{c} > 0$ is a given constant. This equation is a linear first-order partial differential equation. The main goal of this book is to approximate the solution T of this partial differential equation using radial basis functions. Then the approximation itself is a Lyapunov function, and thus can be used to determine the basin of attraction.

Since the function T is not defined at x_0, we also study a second class of Lyapunov functions V which are defined and smooth at x_0. They satisfy the equation $V'(x) = -p(x)$, where $p(x)$ is a given function with certain properties, in particular $p(x_0) = 0$.

For the approximation we use radial basis functions, a powerful meshless approximation method. Given a grid in $\mathbb{R}^n$, the method uses an ansatz for the approximation, such that at each grid point the linear partial differential equation is satisfied. For the other points we derive an error estimate in terms of the grid density.

My Habilitation thesis [21] and the lecture “Basins of Attraction of Dynamical Systems and Algorithms for their Determination” which I held in the winter term 2003/2004 at the University of Technology München were the foundations for this book. I would like to thank J. Scheurle for his support and for many valuable comments. For their support and interest in my work I further wish to thank P. Kloeden, R. Schaback, and H. Wendland. Special thanks to A. Iske who introduced me to radial basis functions and to F. Rupp for his support for the exercise classes to my lecture. Finally, I would like to thank my wife Nicole for her understanding and encouragement during the time I wrote this book.

December 2006 *Peter Giesl*

Contents

Appendices

1

Introduction

1.1 An Example: Chemostat

Let us illustrate our method by applying it to an example. Consider the following situation: a vessel is filled with a liquid containing a nutrient and bacteria, the respective concentrations at time t are given by $x(t)$ and $y(t)$. This family of models is called chemostat, cf. [56]. More generally, a chemostat can also serve as a model for population dynamics: here, $x(t)$ denotes the amount of the prey and $y(t)$ the amount of the predator, e.g. rabbits and foxes.

The vessel is filled with the nutrient at constant rate 1 and the mixture leaves the vessel at the same rate. Thus, the volume in the vessel remains constant. Finally, the bacteria y consumes the nutrient x (or the predator eats the prey), i.e. y increases while x decreases.

The situation is thus described by the following scheme for the temporal rates of change of the concentrations x and y:

- x (nutrient): rate of change = input − washout − consumption
- y (bacteria): rate of change = − washout + consumption

The rates of change lead to the following system of ordinary differential equations, where the dot denotes the temporal derivative: $\dot{} = \frac{d}{dt}$.

$$\begin{cases} \dot{x} = 1 - x - a(x)\,y \\ \dot{y} = -y + a(x)\,y. \end{cases} \tag{1.1}$$

The higher the concentration of bacteria y is, the more consumption takes place. The dependency of the consumption term on the nutrient x is modelled by the non-monotone uptake function $a(x) = \frac{x}{\frac{1}{16}+\frac{x}{4}+x^2}$, i.e. a high concentration of the nutrient has an inhibitory effect. The solution of such a system of differential equations is unique, if the initial concentrations of nutrient and bacteria, $x(0)$ and $y(0)$, respectively, are known at time $t = 0$.

Imagine the right-hand side of the differential equation (1.1) as a vector field $f(x,y) = \begin{pmatrix} 1 - x - a(x)\,y \\ -y + a(x)\,y \end{pmatrix}$. At each point (x,y) the arrows indicate the

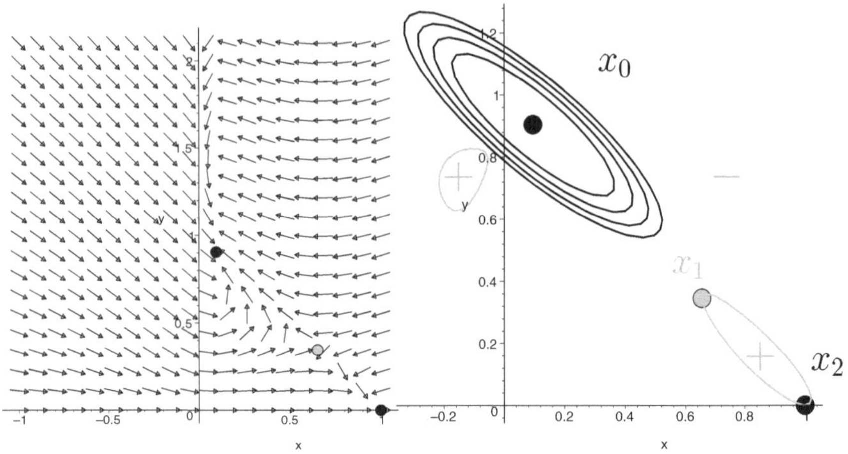

Fig. 1.1. Left: the vector field $f(x, y)$ (arrows with normalized length) and the three equilibria x_1 (unstable, grey), x_0 and x_2 (asymptotically stable, black). Right: the three equilibria and the local Lyapunov function $\mathfrak{v}$: the sign of the orbital derivative $\mathfrak{v}'(x, y)$ (grey) and the level sets $\mathfrak{v}(x, y) = 0.025, 0.02, 0.015, 0.1$ (black) which are ellipses. The sublevel sets are subsets of the basin of attraction $A(x_0)$.

infinitesimal rate of change, to which the solution is tangential, cf. Figure 1.1, left. The norm of the vectors describes the velocity of solutions; note that in Figure 1.1, left, the arrows have normalized length one.

Negative concentrations have no meaning in this model. This is reflected in the equations: solutions starting in the set $S = \{(x, y) \mid x, y \geq 0\}$ do not leave this set in the future, because the vector field at the boundary of S points inwards, cf. Figure 1.1, left. Thus, the set S is called positively invariant.

Points (x, y) where the velocity of the vector filed is zero, i.e. $f(x, y) = 0$, are called equilibria: starting at these points, one stays there for all positive times. In our example we have the three equilibria $x_0 = \left(\frac{3-\sqrt{5}}{8}, \frac{5+\sqrt{5}}{8}\right)$, $x_1 = \left(\frac{3+\sqrt{5}}{8}, \frac{5-\sqrt{5}}{8}\right)$ and $x_2 = (1, 0)$, cf. Figure 1.1. If the initial concentrations are equal to one of these equilibria, then the concentrations keep being the same. What happens, if the initial concentrations are adjacent to these equilibrium-concentrations?

If all adjacent concentrations approach the equilibrium-concentration for $t \to \infty$, then the equilibrium is called asymptotically stable. If they tend away from the equilibrium-concentration, then the equilibrium is called unstable. In the example, x_1 is unstable (grey point in Figure 1.1), while x_0 and x_2 are asymptotically stable (black points in Figure 1.1). The stability of equilibria can often be checked by linearization, i.e. by studying the Jacobian matrix $Df(x_0)$. We know that solutions with initial concentrations near the asymptotically stable equilibrium x_0 tend to x_0. But what does "near" mean?

The set of all initial conditions such that solutions tend to the equilibrium x_0 for $t \to \infty$ is called the basin of attraction $A(x_0)$ of x_0. We are interested in the determination of the basin of attraction. In our example $A(x_0)$ describes the set of initial concentrations so that the concentrations of nutrient and bacteria tend to x_0, which implies that the concentration of the bacteria tends to a constant positive value. If, however, our initial concentrations are in $A(x_2)$, then solutions tend to x_2, i.e. the bacteria will eventually die out.

The determination of the basin of attraction is achieved by a Lyapunov function. A Lyapunov function is a scalar-valued function which decreases along solutions of the differential equation. This can be verified by checking that the orbital derivative, i.e. the derivative along solutions, is negative. One can imagine the Lyapunov function to be a height function, such that solutions move downwards, cf. Figure 1.2. The Lyapunov function enables us to determine a subset K of the basin of attraction by its sublevel sets. These sublevel sets are also positively invariant, i.e. solutions do not leave them in positive time.

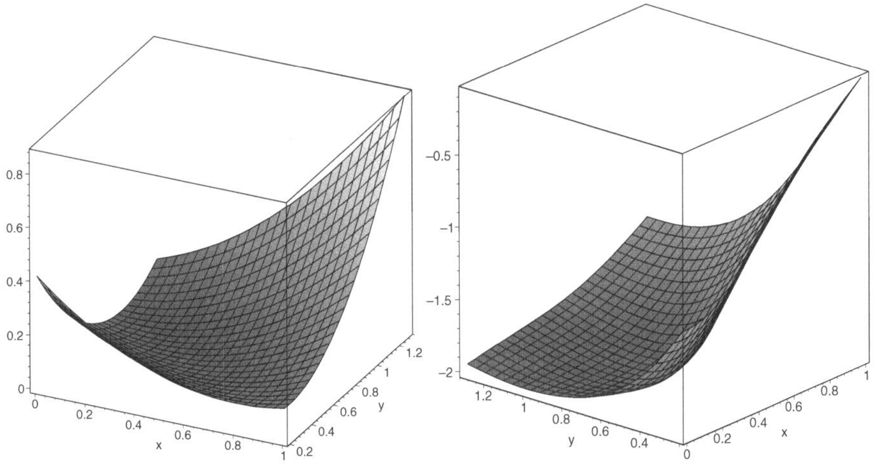

Fig. 1.2. Left: a plot of the local Lyapunov function $\mathfrak{v}$. Note that $\mathfrak{v}$ is a quadratic form. Right: A plot of the calculated Lyapunov function v.

Unfortunately, there is no general construction method for Lyapunov functions. Locally, i.e. in a neighborhood of the equilibrium, a local Lyapunov function can be calculated using the linearization of the vector field f. The orbital derivative of this local Lyapunov function, however, is only negative in a small neighborhood of the origin in general. Figure 1.2, left, shows the local Lyapunov function $\mathfrak{v}(x) = (x - x_0)^T \begin{pmatrix} \frac{1}{2} & \frac{1}{2} \\ \frac{1}{2} & \frac{4}{5}\sqrt{5} - \frac{11}{10} \end{pmatrix} (x - x_0)$, for the determination of $\mathfrak{v}$ cf. Section 2.2.2. In Figure 1.1, right, we see that the orbital

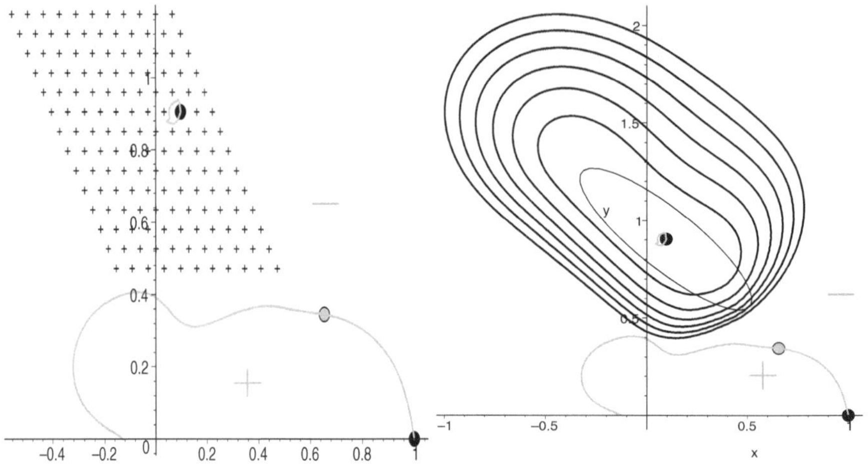

Fig. 1.3. Left: the 153 grid points (black +) for the approximation using radial basis functions and the sign of the orbital derivative $v'(x, y)$ (grey), where v is the calculated Lyapunov function using radial basis functions. Right: the sign of the orbital derivative $v'(x, y)$ (grey), level sets $v(x, y) = -1.7, -1.75, -1.8, -1.85, -1.9, -1.95$, where v is the calculated Lyapunov function using radial basis functions (black), and the sublevel set $\mathfrak{v}(x, y) \leq 0.025$ of the local Lyapunov function (thin black ellipse). This sublevel set covers the points where $v'(x, y) \geq 0$. Hence, sublevel sets of the calculated Lyapunov function are subsets of the basin of attraction $A(x_0)$.

derivative $\mathfrak{v}'$ is negative near x_0 (grey) and thus sublevel sets of $\mathfrak{v}$ (black) are subsets of the basin of attraction.

In this book we will present a method to construct a Lyapunov function in order to determine larger subsets of the basin of attraction. Figure 1.2, right, shows such a calculated Lyapunov function v. In Figure 1.3, right, we see the sign of the orbital derivative $v'(x)$ and several sublevel sets of v. Figure 1.4, left, compares the largest sublevel sets of the local and the calculated Lyapunov function.

The idea of the method evolves from a particular Lyapunov function. Although the explicit construction of a Lyapunov function is difficult, there are theorems available which prove their existence. These converse theorems use the solution of the differential equation to construct a Lyapunov function and since the solutions are not known in general, these methods do not serve to explicitly calculate a Lyapunov function. However, they play an important role for our method.

We study Lyapunov functions fulfilling equations for their orbital derivatives, e.g. the Lyapunov function V satisfying $V'(x) = -\|x - x_0\|^2$. Here, V' denotes the orbital derivative, which is given by $V'(x) = \sum_{i=1}^{2} f_i(x) \frac{\partial V}{\partial x_i}(x)$. Hence, V is the solution of a first-order partial differential equation. We approximate the solution V using radial basis functions and obtain the

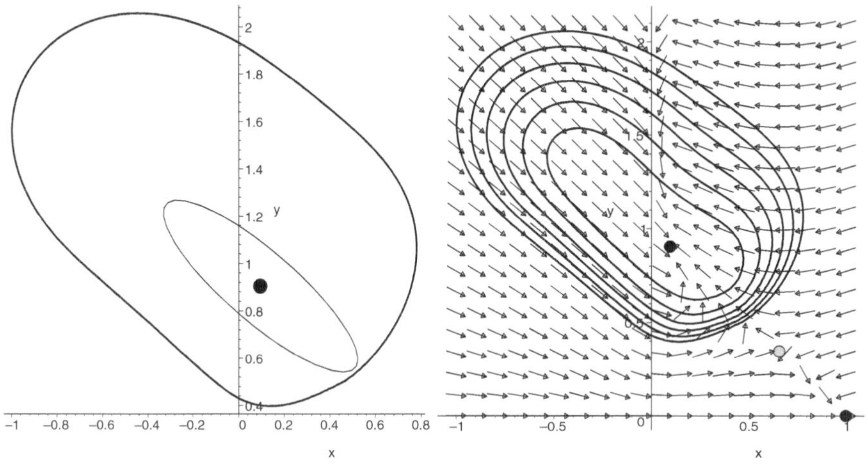

Fig. 1.4. Left: comparison of two subsets of the basin of attraction obtained by the local Lyapunov function (thin black) and by the calculated Lyapunov function (black). Right: the vector field and subsets of the basin of attraction obtained by the calculated Lyapunov function. The subsets are positively invariant – the vector field points inwards.

approximation v. Error estimates for the orbital derivative ensure $v'(x) < 0$ and, thus, the approximation v is a Lyapunov function.

For the radial basis function approximation, we fix a radial basis function $\Psi(x)$ and a grid of scattered points $X_N = \{x_1, \ldots, x_N\}$. In this book we choose the Wendland function family $\psi_{l,k}$ to define the radial basis function by $\Psi(x) = \psi_{l,k}(c\|x\|)$. We use a certain ansatz for the approximating function v and choose the coefficients such that v satisfies the partial differential equation $v'(x_j) = -\|x_j - x_0\|^2$ for all points $x_j \in X_N$ of the grid.

Figure 1.3, left, shows the grid points (black +) that were used for the calculation of v. The sign of v' is negative at the grid points because of the ansatz and also between them due to the error estimate. However, $v'(x)$ is positive near the equilibrium x_0, but this area is covered by the local Lyapunov basin, cf. Figure 1.3, right. Thus, sublevel sets of v are subsets of the basin of attraction. Figure 1.4, left, shows that the calculated Lyapunov function v determines a larger subset of the basin of attraction than the local Lyapunov function $\mathfrak{v}$. All these sublevel sets are subsets of the basin of attraction and, moreover, they are positively invariant, i.e. the vector field at the level sets points inwards, cf. Figure 1.4, right.

Hence, concerning our chemostat example, we have determined subsets of the basin of attraction of x_0, cf. Figure 1.4. If the initial concentrations in the vessel lie in such a set, then solutions tend to the equilibrium x_0 and the bacteria do not die out.

For a similar example, cf. [24], where we also consider a chemostat example, but with a different non-monotone uptake function $a(x)$.

1.2 Lyapunov Functions and Radial Basis Functions

In this section we review the literature on Lyapunov functions and radial basis functions

Lyapunov Functions

The literature on Lyapunov functions is very large; for an overview cf. Hahn [34]. In 1893, Lyapunov [48] introduced his direct or second method, where he sought to obtain results concerning the stability of an equilibrium without knowing the solution of the differential equation, but by only using the differential equation itself. He used what later was called Lyapunov functions and proved that a strict Lyapunov function implies the asymptotic stability of the equilibrium. Barbašin and Krasovskiĭ [6] showed that the basin of attraction is the whole phase space if the Lyapunov function is radially unbounded. Hahn describes how a Lyapunov function can be used to obtain a subset of the basin of attraction through sublevel sets, cf. [35] pp. 108/109 and 156/157.

Converse theorems which guarantee the existence of such a Lyapunov function under certain conditions have been given by many authors, for an overview cf. [35] or [58]. The first main converse theorem for asymptotic stability was given by Massera [50] in 1949 and it was improved by many authors in several directions. However, all the existence theorems offer no method to explicitly construct Lyapunov functions.

Krasovskiĭ writes in 1959: "One could hope that a method for proving the existence of a Lyapunov function might carry with it a constructive method for obtaining this function. This hope has not been realized", [46], pp. 11/12. He suggests [46], p. 11, to start from a given system, find a simpler system which approximates the original one and for which one can show stability, and then to prove that the corresponding property also holds for the original system.

For linear systems one can construct a quadratic Lyapunov function of the form $\mathfrak{v}(x) = (x - x_0)^T B(x - x_0)$ with a symmetric, positive definite matrix B, where x_0 denotes the equilibrium, cf. e.g. [33]. In [34], pp. 29/30, Hahn describes, starting from a nonlinear system, how to use the quadratic Lyapunov function of the linearized system as a Lyapunov function for the nonlinear system. He also discusses the search for a sublevel set inside the region $\mathfrak{v}'(x) < 0$, which is a subset of the basin of attraction.

Many approaches consider special Lyapunov functions like quadratic, polynomial, piecewise linear, piecewise quadratic or polyhedral ones, which are special piecewise linear functions. Often these methods can only be applied to special differential equations.

Piecewise linear functions are particularly appropriate for the implementation on computers since they only depend on a finite number of values. Julián [42] approximated the differential equation by a piecewise linear right-hand side and constructed a piecewise linear Lyapunov function using linear programming (linear optimization). Hafstein (formerly Marinossón), cf. [49] or [32], improved this ansatz and constructed a piecewise linear Lyapunov function for the original nonlinear system also using linear programming. Moreover, he included an error analysis in his ansatz. On the other hand he could not guarantee that the subsets which he determines with his method cover the whole basin of attraction. In some of his examples the calculated subset is even smaller than the one obtained by the Lyapunov function for the linearized system with a sharper estimate.

A different method deals with the Zubov equation and computes a solution of this partial differential equation. Since the solution of the Zubov equation determines the whole basin of attraction, one can cover each compact subset of the basin of attraction with an approximate solution. For computational aspects, cf. e.g. Genesio et al. [19]. In a similar approach to Zubov's method, Vannelli & Vidyasagar [59] use a rational function as Lyapunov function candidate and present an algorithm to obtain a maximal Lyapunov function in the case that f is analytic.

In Camilli et al. [12], Zubov's method was extended to control problems in order to determine the robust domain of attraction. The corresponding generalized Zubov equation is a Hamilton-Jacobi-Bellmann equation. This equation has a viscosity solution which can be approximated using standard techniques after regularization at the equilibrium, e.g. one can use piecewise affine approximating functions and adaptive grid techniques, cf. Grüne [30] or Camilli et al. [12]. The method works also for non-smooth f since the solution is not necessarily smooth either. The error estimate here is given in terms of $|v_\epsilon(x) - \tilde{v}_\epsilon(x)|$, where v_ϵ denotes the regularized Lyapunov function and $\tilde{v}_\epsilon$ its approximation, and not in terms of the orbital derivative.

In this book we present a new method to construct Lyapunov functions. We start from a converse theorem proving the existence of a Lyapunov function T which satisfies the equation $T'(x) = -\bar{c}$, where $\bar{c} > 0$ is a given constant. This equation is a linear first-order partial differential equation due to the formula for the orbital derivative:

$$T'(x) = \sum_{i=0}^{n} f_i(x) \frac{\partial T}{\partial x_i}(x) = -\bar{c}. \tag{1.2}$$

The main goal of this book is to approximate the solution T of (1.2) by a function t using radial basis functions. It turns out that t itself is a Lyapunov function, i.e. $t'(x)$ is negative, and thus can be used to determine the basin of attraction. The approximation error will be estimated in terms of $|T'(x) - t'(x)| \leq \iota$. Hence, $t'(x) \leq T'(x) + \iota = -\bar{c} + \iota < 0$ if the error $\iota < \bar{c}$ is small enough.

However, the function T is not defined at x_0. Hence, we consider a second class of Lyapunov functions V which are defined and smooth at x_0. They satisfy the equation $V'(x) = -p(x)$, where $p(x)$ is a given function with certain properties, in particular $p(x_0) = 0$, and we often use $p(x) = \|x - x_0\|^2$. The equation $V'(x) = -p(x)$ is a modified Zubov equation, cf. [24]. In the following we denote by Q one of these Lyapunov function of type $Q = T$ or $Q = V$. For the approximation of Q we use radial basis functions.

Radial Basis Functions

Radial basis functions are a powerful tool to solve partial differential equations. For an overview cf. [63], [11], [10], or [52], for a tutorial cf. [40]. The main advantage of this method is that it is meshless, i.e. no triangulation of the space $\mathbb{R}^n$ is needed. Other methods, e.g. finite element methods, first generate a triangulation of the space, use functions on each part of the triangulation, e.g. affine functions as in some examples discussed above, and then patch them together obtaining a global function. The resulting function is not very smooth and the method is not very effective in higher space dimensions. Moreover, the interpolation problem stated by radial basis functions is always uniquely solvable. Radial basis functions give smooth approximations, but at the same time require smooth functions that are approximated. In our applications, as we will see, the freedom of choosing the grid in an almost arbitrary way will be very advantageous.

Let us explain the approximation with radial basis functions. Denote by D the linear operator of the orbital derivative, i.e. $DQ(x) = Q'(x)$. We use the symmetric ansatz leading to a symmetric interpolation matrix A. One defines a grid $X_N = \{x_1, \ldots, x_N\} \subset \mathbb{R}^n$. The reconstruction (approximation) q of the function Q is obtained by the ansatz $q(x) = \sum_{k=1}^N \beta_k \langle \nabla_y \Psi(x-y)\big|_{y=x_k}, f(x_k)\rangle$ with coefficients $\beta_k \in \mathbb{R}$. The function Ψ is the radial basis function. In this book we use $\Psi(x) = \psi_{l,k}(c\|x\|)$ where $\psi_{l,k}$ is a Wendland function. Wendland functions are positive definite functions (and not only conditionally positive definite) and have compact support. The coefficients β_k are determined by the claim that $q'(x_j) = Q'(x_j)$ holds for all grid points $j = 1, \ldots, N$. This is equivalent to a system of linear equations $A\beta = \alpha$ where the interpolation matrix A and the right-hand side vector α are determined by the grid and the values $Q'(x_j)$. The interpolation matrix A is a symmetric $(N \times N)$ matrix, where N is the number of grid points. We show that A is positive definite and thus the linear equation has a unique solution β. Provided that Q is smooth enough, one obtains an error estimate on $|Q'(x) - q'(x)|$ depending on the density of the grid.

While the interpolation of function values has been studied in detail since the 1970s, the interpolation via the values of a linear operator and thus the solutions of PDEs has only been considered since the 1990s. The values of such linear operators are also called Hermite-Birkhoff data. They have been studied, e.g. by Iske [38], Wu [67], Franke & Schaback [17] and [18] and Wendland [63]. Franke & Schaback approximated the solution of a Cauchy problem in partial differential equations, cf. also [54]. This results in a mixed problem, combining different linear operators, cf. [17] and [18]. Their error estimates used the fact that the linear operator is translation invariant. The partial differential equations they studied thus have constant coefficients. Our linear operator D, however, is the orbital derivative and is not translation invariant. Error estimates hence have to use different techniques at a certain point, which are Taylor expansions in our case.

1.3 Overview

Chapter 2 deals with the dynamical systems' part of the book. After an introduction and the definition of a Lyapunov function and a Lyapunov basin (sublevel set), we show in Theorem 2.24 that a Lyapunov basin is a subset of the basin of attraction. The local Lyapunov functions $\mathfrak{d}$ and $\mathfrak{v}$ are constructed and discussed. Then the global Lyapunov functions T and V satisfying $T'(x) = -\bar{c}$ and $V'(x) = -p(x)$ are considered. In Section 2.3.2 general properties of Lyapunov functions and their level sets are discussed. In Section 2.3.4 the Taylor polynomial of V is constructed and its properties are shown. The chapter closes with a summary and examples of the different Lyapunov functions.

Chapter 3 deals with the radial basis functions' part of the book. We discuss the approximation of a function Q via the function values $Q(x)$, via its orbital derivative $DQ(x) = Q'(x)$ and via the values of the operator $D_m Q(x) = Q'(x) + m(x)Q(x)$ where m is a scalar-valued function. Moreover, the mixed approximation is studied, where the orbital derivative $Q'(x)$ is given on a grid X_N and the function values $Q(x)$ are given on a second grid X_M^0. In Section 3.1.4 the Wendland functions, cf. [62], a certain class of radial basis functions with compact support, are introduced. In Section 3.2 the native space and its dual space are defined. They are Sobolev spaces in the case of the Wendland functions. We show that the interpolation matrices are positive definite and obtain error estimates.

In Chapter 4 we combine the preceding chapters to construct a Lyapunov function. We approximate the two global Lyapunov functions T and V by the approximating functions t and v using radial basis functions. We show that $t'(x)$ and $v'(x)$ are negative for $x \in K \setminus U$ where $K \subset A(x_0)$ is a compact set and U is a neighborhood of x_0, provided that the grid of the radial basis function approximation is dense enough. Since the approximating functions can have positive orbital derivative in the neighborhood U of x_0 this approximation is called *non-local*. For the *local part* we present three methods: the non-local Lyapunov basin can be combined with a local Lyapunov basin. Another option is the combination of the non-local Lyapunov function with a local Lyapunov function by a partition of unity. The third possibility is to approximate V via an approximation of a certain function W using the Taylor polynomial of V.

In Chapter 5 we show that this method enables us to determine the whole basin of attraction. In particular we show that all compact subsets of the basin of attraction can be obtained with our method. This is true for the approximation of V – either directly or using its Taylor polynomial. If we use a mixed approximation, then we also obtain the whole basin of attraction by approximating T. Moreover, the mixed approximation is particularly appropriate to exhaust the basin of attraction step by step.

In Chapter 6 we describe three possibilities to use our method for the actual computation of the basin of attraction and illustrate them with examples. In an appendix we provide basic facts and some special results from distribution theory. Moreover, the data used for the computation of all figures is listed.

2

Lyapunov Functions

2.1 Introduction to Dynamical Systems

In this section we give a short introduction to dynamical systems. For a text-book, cf. [29], [60], [2], [5] or [66]. We summarize the important definitions and concepts which we will need in the sequel. In particular, we define equilibria, their stability and their basin of attraction. Moreover, we give the definition of a Lyapunov function with Lyapunov basin. The most important results of this section for the sequel are Theorems 2.24 and 2.26, where we prove that a Lyapunov basin is a subset of the basin of attraction.

2.1.1 Basic Definitions and Concepts

Throughout the book we consider the autonomous system of differential equations

$$\dot{x} = f(x), \tag{2.1}$$

where $f \in C^\sigma(\mathbb{R}^n, \mathbb{R}^n)$, $\sigma \geq 1$, $n \in \mathbb{N}$. The initial value problem $\dot{x} = f(x)$, $x(0) = \xi$ has a unique solution $x(t)$ for all initial values $\xi \in \mathbb{R}^n$. The solution $x(t)$ depends continuously on the initial value ξ and exists locally in time, i.e. for $t \in I$, where $0 \in I \subset \mathbb{R}$ is an open interval, cf. e.g. [2] or [5]. Hence, we can define the flow operator.

Definition 2.1 (Flow operator). *Define the operator S_t by $S_t\xi := x(t)$, where $x(t)$ is the solution of the initial value problem $\dot{x} = f(x)$, $x(0) = \xi \in \mathbb{R}^n$ for all $t \in \mathbb{R}$ for which this solution exists.*

An abstract (continuous) dynamical system is defined in the following way.

Definition 2.2 (Dynamical System). *Let X be a complete metric space. $(X, \mathbb{R}_0^+, S_t)$ is a continuous dynamical system if $S_t \colon X \to X$ is defined for* *all $t \in \mathbb{R}_0^+$, $(t, x) \mapsto S_t x$ is a continuous mapping with respect to both x and t and, moreover, S_t is a semigroup, i.e. $S_0 = \mathrm{id}$ and $S_{t+s} = S_t \circ S_s$ for all $t, s \in \mathbb{R}_0^+$.*

We will consider $X = \mathbb{R}^n$ with the Euclidean norm $\|\cdot\|$. As we mentioned above, existence, uniqueness and continuous dependence of the flow $S_t x$ on x and t follow by the smoothness of f, cf. e.g. [36]. However, in general the existence of solutions is only true for a local time interval I. If the existence is ensured for all positive times, then the flow operator S_t defines a dynamical system.

Remark 2.3 *Assume that for all $\xi \in \mathbb{R}^n$ the solution $x(t)$ of the initial value problem $\dot{x} = f(x)$, $x(0) = \xi$ exists for all $t \in \mathbb{R}_0^+$. Then $(\mathbb{R}^n, \mathbb{R}_0^+, S_t)$ where S_t is the flow operator of Definition 2.1, defines a dynamical system in the sense of Definition 2.2.*

Although many of the following definitions and results hold for general dynamical systems as in Definition 2.2, we restrict ourselves from now on to dynamical systems given by the differential equation (2.1).

Definition 2.4 (Positive orbit). *Let $(\mathbb{R}^n, \mathbb{R}_0^+, S_t)$ be a dynamical system. The positive orbit of $x \in \mathbb{R}^n$ is defined by $O^+(x) = \bigcup_{t \geq 0} S_t x \subset \mathbb{R}^n$.*

We can ensure the existence for all $t \in \mathbb{R}$ by considering a slightly different differential equation.

Remark 2.5 *Consider the system of differential equations:*

$$\dot{x} = \frac{f(x)}{1 + \|f(x)\|^2} =: g(x) \tag{2.2}$$

Note that $g \in C^\sigma(\mathbb{R}^n, \mathbb{R}^n)$ if $f \in C^\sigma(\mathbb{R}^n, \mathbb{R}^n)$.

The solutions of (2.1) and (2.2) have the same positive orbits since g is obtained by multiplication of f with a scalar, positive factor. Hence, dynamically the two systems have the same properties. Positive orbits of both dynamical systems are the same, only the velocity is different. Since $\|g(x)\| \leq \frac{1}{2}$, solutions of (2.2) exist for all $t \in \mathbb{R}$.

From now on we assume that (2.1) defines a dynamical system. In the following definition we define equilibria which represent the simplest solutions of (2.1).

Definition 2.6 (Equilibrium). *A point $x_0 \in \mathbb{R}^n$ is called an* equilibrium *of (2.1), if $f(x_0) = 0$.*

If x_0 is an equilibrium, then $S_t x_0 = x_0$ for all $t \geq 0$, i.e. $x(t) = x_0$ is a constant solution.

In the next definition we provide terms to discuss the behavior of solutions near equilibria. In particular, an equilibrium is called stable if adjacent solutions stay adjacent to the equilibrium. An equilibrium is called attractive, if adjacent solutions converge to the equilibrium as time tends to infinity.

Stability and attractivity are different properties of an equilibrium, neither implies the other. We call an equilibrium *asymptotically stable* if it is both stable and attractive. It is called exponentially asymptotically stable if the rate of convergence is exponential. We will later assume, that the system (2.1) has an exponentially asymptotically stable equilibrium.

Definition 2.7 (Stability and attractivity). *Let x_0 be an equilibrium.*

- x_0 *is called* stable, *if for all $\epsilon > 0$ there is a $\delta > 0$ such that $S_t x \in B_\epsilon(x_0) := \{x \in \mathbb{R}^n \mid \|x - x_0\| < \epsilon\}$ holds for all $x \in B_\delta(x_0)$ and all $t \geq 0$. Here and below, $\|\cdot\|$ denotes the Euclidean norm in $\mathbb{R}^n$.*
- x_0 *is called* unstable *if x_0 is not stable.*
- x_0 *is called* attractive, *if there is a $\delta' > 0$ such that $\|S_t x - x_0\| \stackrel{t\to\infty}{\longrightarrow} 0$ holds for all $x \in B_{\delta'}(x_0)$.*
- x_0 *is called* asymptotically stable, *if x_0 is stable and attractive.*
- x_0 attracts a set $K \subset \mathbb{R}^n$ uniformly, *if* $\operatorname{dist}(S_t K, x_0) := \sup_{x\in K} \|S_t x - x_0\| \stackrel{t\to\infty}{\longrightarrow} 0$.
- x_0 *is called* exponentially asymptotically stable *(with exponent $-\nu < 0$), if x_0 is stable and there is a $\delta' > 0$ such that $\|S_t x - x_0\| e^{+\nu t} \stackrel{t\to\infty}{\longrightarrow} 0$ holds for all $x \in B_{\delta'}(x_0)$.*

Let is consider a linear right-hand side $f(x) = Ax$ where A denotes an $(n \times n)$ matrix. For these linear differential equations

$$\dot{x} = Ax$$

the solution of the initial value problem $\dot{x} = Ax$, $x(0) = \xi$ is given by $x(t) = \exp(At) \cdot \xi$. Thus, in the linear case the solutions are known explicitly. From the formula for the solutions it is clear that the stability and attractivity of the equilibrium $x_0 = 0$ depends on the real parts of the eigenvalues of A. More precisely, the origin is stable if and only if the real parts of all eigenvalues are non-positive and the geometric multiplicity equals the algebraic multiplicity for all eigenvalues with vanishing real part. The origin is asymptotically stable if the real parts of all eigenvalues are negative. In this case, the origin is even exponentially asymptotically stable.

Now we return to a general nonlinear differential equation $\dot{x} = f(x)$. The stability of an equilibrium x_0 can often be studied by the linearization around x_0, i.e. by considering the linear system $\dot{x} = Df(x_0)(x - x_0)$. The real parts of the eigenvalues of $Df(x_0)$ provide information about the stability of the linear system. If all eigenvalues have strictly negative real part, then the equilibrium x_0 is (exponentially) asymptotically stable – not only with respect to the linearized but also with respect to the original nonlinear system. While this condition on the real parts of the eigenvalues is only sufficient for asymptotical stability of the nonlinear system, it is sufficient and necessary for exponential asymptotic stability. For a proof of the following well-known proposition, cf. e.g. [36].

Proposition 2.8 *Let x_0 be an equilibrium of (2.1). x_0 is exponentially asymptotically stable, if and only if* $\text{Re}\,\lambda < 0$ *holds for all eigenvalues λ of $Df(x_0)$.*

For an asymptotically stable equilibrium x_0 there is a neighborhood $B_{\delta'}(x_0)$ which is attracted by x_0 for $t \to \infty$, cf. Definition 2.7. The set of *all* initial points of $\mathbb{R}^n$, the solutions of which eventually tend to x_0, is called the basin of attraction of x_0.

Definition 2.9 (Basin of attraction). *The* basin of attraction *of an asymptotically stable equilibrium x_0 is defined by*

$$A(x_0) := \{x \in \mathbb{R}^n \mid S_t x \stackrel{t\to\infty}{\longrightarrow} x_0\}.$$

Example 2.10 *We give several examples.*

1. *Consider the linear system $\dot{x} = -x$, $x \in \mathbb{R}$. The solution of the initial value problem $x(0) = \xi$ is given by $x(t) = e^{-t}\xi$. 0 is an exponentially asymptotically stable equilibrium with basin of attraction $A(0) = \mathbb{R}$.*
2. *We study the nonlinear system $\dot{x} = \frac{1}{1+x^4}x(x^2-1) =: f(x)$, $x \in \mathbb{R}$. The system has three equilibria $x_0 = 0$, $x_1 = 1$, $x_2 = -1$. $x_0 = 0$ is exponentially asymptotically stable since $f'(0) = -1$. The other two equilibria are unstable. The basin of attraction of 0 is $A(0) = (-1, 1)$.*
3. *Consider the two-dimensional nonlinear system*

$$\begin{cases} \dot{x} = -x(1-x^2-y^2) - y \\ \dot{y} = -y(1-x^2-y^2) + x \end{cases}$$

In polar coordinates (r, φ), where $x = r\cos\varphi$, $y = r\sin\varphi$, the system reads $\dot{r} = -r(1-r^2)$, $\dot{\varphi} = 1$. The only equilibrium is the origin, it is exponentially asymptotically stable since $Df(0,0) = \begin{pmatrix} -1 & -1 \\ 1 & -1 \end{pmatrix}$ and the eigenvalues are $-1 \pm i$, i.e. both have negative real part. The basin of attraction of the origin is given by the unit disc $A(0,0) = \{(x,y) \in \mathbb{R}^2 \mid x^2 + y^2 < 1\}$.
The unit sphere in this example is a periodic orbit, i.e. a solution $x(t)$ such that $x(0) = x(T)$ with $T > 0$.

Proposition 2.11 (Properties of the basin of attraction) *Let x_0 be an asymptotically stable equilibrium. Then the basin of attraction $A(x_0)$ is non-empty and open.*

Let $K \subset A(x_0)$ be a compact set. Then x_0 attracts K uniformly.

PROOF: x_0 is an element of the basin of attraction, which is thus non-empty by definition. Let $B_{\delta'}(x_0)$ be as in Definition 2.7 and let $x \in A(x_0)$. By definition there is a T_0 such that $S_{T_0}x \in B_{\frac{\delta'}{2}}(x_0)$. Since S_{T_0} is continuous, there is a $\delta > 0$ such that $S_{T_0}B_\delta(x) \subset B_{\frac{\delta'}{2}}(S_{T_0}x) \subset B_{\delta'}(x_0)$. Since $B_{\delta'}(x_0)$ is attracted by x_0, $S_{T_0}B_\delta(x) \in A(x_0)$ and thus also $B_\delta(x) \subset A(x_0)$; in particular, $A(x_0)$ is open.

We prove the uniform attractivity: Assuming the opposite, there is an $\epsilon > 0$ and sequences $t_k \to \infty$ and $x_k \in K$, such that $S_{t_k}x_k \notin B_\epsilon(x_0)$ holds for all $k \in \mathbb{N}$. Since x_0 is stable, there is a $\delta > 0$ such that $x \in B_\delta(x_0)$ implies $S_t x \in B_\epsilon(x_0)$ for all $t \geq 0$.

Since K is compact there is a convergent subsequence of $x_k \in K$, which we still denote by x_k, with limit $x \in K$. Since $x \in K \subset A(x_0)$, there is a $T_0 \geq 0$ with $S_{T_0}x \in B_{\frac{\delta}{2}}(x_0)$. Since S_{T_0} is continuous, there is a $\tilde{\delta} > 0$, such that $y \in B_{\tilde{\delta}}(x)$ implies $S_{T_0}y \in B_{\frac{\delta}{2}}(S_{T_0}x)$. Now let N be so large that both $t_N \geq T_0$ and $x_N \in B_{\tilde{\delta}}(x)$ hold. Then $S_{T_0}x_N \in B_{\frac{\delta}{2}}(S_{T_0}x) \subset B_\delta(x_0)$ and hence $S_{t_N}x_N \in B_\epsilon(x_0)$, which is a contradiction. □

The long-time behavior of solutions is characterized by the ω-limit set.

Definition 2.12 (ω-limit set). *We define the* ω-limit set *for a point* $x \in \mathbb{R}^n$ *with respect to (2.1) by*

$$\omega(x) := \bigcap_{s \geq 0} \overline{\bigcup_{t \geq s} S_t x}. \tag{2.3}$$

We have the following characterization of limit sets: $w \in \omega(x)$ *if and only if there is a sequence* $t_k \stackrel{k\to\infty}{\longrightarrow} \infty$ *such that* $S_{t_k}x \stackrel{k\to\infty}{\longrightarrow} w$.

Definition 2.13 (Positively invariant). *A set* $K \subset \mathbb{R}^n$ *is called* positively invariant *if* $S_t K \subset K$ *holds for all* $t \geq 0$.

Lemma 2.14. *Let* $K \subset \mathbb{R}^n$ *be a compact, positively invariant set. Then* $\varnothing \neq \omega(x) \subset K$ *for all* $x \in K$.

PROOF: Let $x \in K$. Since K is positively invariant, $S_t x \in K$ for all $t \geq 0$. Since K is compact, the limit of any convergent sequence $S_{t_k}x$ lies in K and thus $\omega(x) \subset K$. Moreover, the sequence $S_n x$, $n \in \mathbb{N}$ has a convergent subsequence and hence $\omega(x) \neq \varnothing$. □

Example 2.15 *For Example 2.10, 2. we have* $\omega(x) = \{0\}$ *for all* $x \in (-1,1)$, $\omega(x) = \varnothing$ *for all* $x \in (-\infty,-1) \cup (1,\infty)$ *and* $\omega(1) = \{1\}$, $\omega(-1) = \{-1\}$. $K_1 = \{0\}$, $K_2 = (-1/2, 1/2)$, $K_3 = (-1, 1]$, $K_4 = (0,1)$ *and* $K_5 = \mathbb{R}$ *are examples for positively invariant sets.*

As a second example we consider the two-dimensional nonlinear system $\begin{cases} \dot{x} = x(1-x^2-y^2) + y \\ \dot{y} = y(1-x^2-y^2) - x \end{cases}$. *This is Example 2.10, 3. after inversion of time.*

In polar coordinates the system reads $\dot{r} = r(1 - r^2)$, $\dot{\varphi} = -1$. *Here we have* $\omega(0,0) = \{(0,0)\}$ *and* $\omega(x,y) = \{(\xi,\eta) \mid \xi^2 + \eta^2 = 1\}$ *for all* $(x,y) \neq (0,0)$, *i.e. the* ω*-limit set of all points except for the origin is the periodic orbit.*

Lemma 2.16. *Let* x_0 *be a stable equilibrium and* $\epsilon > 0$. *Then there is a positively invariant neighborhood* $U \subset B_\epsilon(x_0)$ *of* x_0.

PROOF: Since x_0 is stable, there is a $\delta > 0$ such that $x \in B_\delta(x_0)$ implies $S_t x \in B_\epsilon(x_0)$ for all $t \geq 0$. Set $U := \bigcup_{t\geq 0} S_t B_\delta(x_0) = \{S_t x \mid x \in B_\delta(x_0), t \geq 0\}$. This set is positively invariant by construction and a neighborhood of x_0 since $B_\delta(x_0) \subset U$. Moreover, $U \subset B_\epsilon(x_0)$ by stability of x_0. □

The following Lemma 2.17 will be used for the proof of Theorem 2.24.

Lemma 2.17. *Let* x_0 *be a stable equilibrium and let* $\omega(x) = \{x_0\}$ *for an* $x \in \mathbb{R}^n$. *Then* $\lim_{t\to\infty} S_t x = x_0$.

PROOF: Assuming the opposite, there is an $\epsilon > 0$ and a sequence $t_k \to \infty$, such that $S_{t_k} x \notin B_\epsilon(x_0)$ holds for all $k \in \mathbb{N}$. Since x_0 is stable, there is a positively invariant neighborhood U of x_0 with $U \subset B_\epsilon(x_0)$ by Lemma 2.16. Since $\omega(x) = \{x_0\}$ there is a $T_0 \geq 0$, such that $S_{T_0} x \in U$. Hence, $S_{T_0+t} x \in U \subset B_\epsilon(x_0)$ for all $t \geq 0$: contradiction to the assumption for all t_k with $t_k \geq T_0$. □

Now we consider a function $Q\colon \mathbb{R}^n \to \mathbb{R}$. We define its orbital derivative which is its derivative along a solution of (2.1).

Definition 2.18 (Orbital derivative). *The* orbital derivative *of a function* $Q \in C^1(\mathbb{R}^n, \mathbb{R})$ *with respect to (2.1) at a point* $x \in \mathbb{R}^n$ *is denoted by* $Q'(x)$ *(in contrast to the derivative with respect to the time t:* $\dot{x}(t) = \frac{d}{dt}x(t)$*). It is defined by*

$$Q'(x) = \langle \nabla Q(x), f(x) \rangle.$$

Remark 2.19 *The orbital derivative is the derivative along a solution: with the chain rule we have*

$$\frac{d}{dt} Q(S_t x)\bigg|_{t=0} = \langle \nabla Q(S_t x), \dot{x}(t) \rangle\bigg|_{t=0} = \langle \nabla Q(x), f(x) \rangle = Q'(x).$$

LaSalle's principle states that for points z in the ω-limit set we have $Q'(z) = 0$.

Theorem 2.20 (LaSalle's principle). *Let* $\varnothing \neq K \subset \mathbb{R}^n$ *be a closed and positively invariant set. Let* $Q \in C^1(\mathbb{R}^n, \mathbb{R})$ *be such that* $Q'(x) \leq 0$ *holds for all* $x \in K$.

Then $z \in \omega(x)$ *for an* $x \in K$ *implies* $Q'(z) = 0$.

PROOF: Let $x \in K$. If $\omega(x) = \varnothing$, then there is nothing to show. Now let $z \in \omega(x)$. Then $z \in K$, since K is closed. Assume in contradiction to the theorem that $Q'(z) \neq 0$, i.e. $Q'(z) < 0$. Then there is a $\tau > 0$ such that $Q(S_\tau z) < Q(z)$. There is a sequence $t_k \to \infty$ with $S_{t_k} x \to z$ and without loss of generality we can assume $t_{k+1} - t_k > \tau$ for all $k \in \mathbb{N}$. Then $Q(S_{t_{k+1}} x) \leq Q(S_{\tau+t_k} x)$ since $Q'(S_t x) \leq 0$ for all $t \geq 0$ by assumption. Hence, $S_{\tau+t_k} x \to S_\tau z$ which implies $Q(S_{\tau+t_k} x) \to Q(S_\tau z)$. This leads to the contradiction $Q(z) \leq Q(S_\tau z)$.

Hence, we have shown $Q'(z) = 0$. □

2.1.2 Lyapunov Functions

The idea of a Lyapunov function can be illustrated by the following example: consider a heavy ball on some mountainous landscape; the gravitational force acts on the ball. The deepest point of a valley corresponds to an asymptotically stable equilibrium: starting at this point, the ball remains there for all times. Starting near the equilibrium in the valley, the ball stays near the equilibrium (stability) and, moreover, approaches the equilibrium as times tends to infinity (attractivity). The basin of attraction consists of all starting points, such that the ball approaches the equilibrium.

A Lyapunov function in this simple example is the height. The classical definition of a Lyapunov function is a function $Q\colon \mathbb{R}^n \to \mathbb{R}$ with (i) a strict local minimum at x_0 which (ii) decreases along solutions. The first property is fulfilled since x_0 is the deepest point of a valley and the second by the gravitation which forces the ball to loose height. For differential equations which model a dissipative physical situation, the energy is a candidate for a Lyapunov function: it decreases along solutions due to the dissipativity of the system. Hence, solutions tend to a local minimum of the energy (i). For concrete examples, the property (ii) can be checked without explicitly calculating the solution: the orbital derivative has to be negative, i.e. $Q'(x) < 0$ for $x \neq x_0$.

We can also use the Lyapunov function to obtain a subset of the basin of attraction: the sublevel set $\{x \in B \mid Q(x) \leq R^2\}$, where B is some neighborhood of x_0, is a subset of the basin of attraction, provided that $Q'(x)$ is negative here. This set is positively invariant, i.e. solutions starting in the set do not leave it in the future.

In this book we give a slightly different definition of a Lyapunov function for an equilibrium x_0. We call a function $Q \in C^1(\mathbb{R}^n, \mathbb{R})$ a *Lyapunov function* if there is a neighborhood K of x_0 with $Q'(x) < 0$ for all $x \in K \setminus \{x_0\}$.

A function $Q \in C^1(\mathbb{R}^n, \mathbb{R})$ and a compact set K with $x_0 \in \mathring{K}$ are called a Lyapunov function Q with *Lyapunov basin* K, if there is an open neighborhood B of K such that the following two conditions are satisfied:

- sublevel set: $K = \{x \in B \mid Q(x) \leq R^2\}$ and
- negative orbital derivative: $Q'(x) < 0$ holds for all $x \in K \setminus \{x_0\}$.

Note that compared to the classical definition we only assume the property (ii) concerning the orbital derivative. The property (i), i.e. Q has a strict local minimum at x_0, turns out to be a consequence of the above definition of a Lyapunov function Q and a Lyapunov basin K, cf. Theorem 2.24. This theorem also states that a Lyapunov basin K is always a subset of the basin of attraction. Hence, the goal of this book will be to find such a pair of Lyapunov function and Lyapunov basin.

The level sets of Lyapunov functions of an exponentially asymptotically stable equilibrium are diffeomorphic to S^{n-1} as we show in Theorem 2.45. Hence, supposed we are given a Lyapunov function with negative orbital derivative in $K \setminus \{x_0\}$ where K is a compact neighborhood of x_0, there is a corresponding Lyapunov basin which reaches up to the boundary of K. Thus, a Lyapunov function for an exponentially asymptotically stable equilibrium always provides a Lyapunov basin and thus a subset of the basin of attraction. The main goal is thus to find a Lyapunov function, i.e. a function with negative orbital derivative.

In 1893, Lyapunov [48] introduced what later was called Lyapunov function. He used these functions for the stability analysis of an equilibrium without knowledge about the solutions of the differential equation, but only using the differential equation itself. Lyapunov [48] proved that a strict (classical) Lyapunov function implies the asymptotic stability of the equilibrium. Barbašin and Krasovskiĭ [6] showed that the basin of attraction is the whole phase space if the Lyapunov function is radially unbounded, cf. also [35], p. 109 and [9], p. 68. Hahn describes in [35], pp. 108/109 and 156/157, how a Lyapunov function can be used to obtain a subset of the basin of attraction.

Classically, a function $Q \in C^1(\mathbb{R}^n, \mathbb{R})$ is called a strict Lyapunov function for an equilibrium x_0 if both

- $Q'(x) < 0$ holds for all $x \in K \setminus \{x_0\}$ and
- $Q(x) > Q(x_0)$ holds for all $x \in K \setminus \{x_0\}$

where K is some neighborhood of x_0. Then x_0 is an asymptotically stable equilibrium. The idea for the proof is that solutions near x_0 decrease along solutions and thus tend to the minimum, which is x_0.

We use the following definition of a Lyapunov function in this book.

Definition 2.21 (Lyapunov function). *Let x_0 be an equilibrium of $\dot{x} = f(x)$, $f \in C^1(\mathbb{R}^n, \mathbb{R}^n)$.*

Let $B \ni x_0$ be an open set. A function $Q \in C^1(B, \mathbb{R})$ is called Lyapunov function *if there is a set $K \subset B$ with $x_0 \in \overset{\circ}{K}$, such that*

$$Q'(x) < 0 \textit{ for all } x \in K \setminus \{x_0\}.$$

Remark 2.22 *If the set K is small, we call Q a local Lyapunov function. If there is a neighborhood $E \subset K$ of x_0 (exceptional set) such that $Q'(x) < 0$ holds for $x \in K \setminus E$, then Q is called a non-local Lyapunov function. Note* *that such a function is not a Lyapunov function in the sense of Definition 2.21. However, if we have additional information on the set E, e.g. by a local Lyapunov function, then we can draw similar conclusions as in the case of a Lyapunov function, cf. Definition 2.25 and Theorem 2.26.*

If, in addition, we assume that the set K in Definition 2.21 is a compact sublevel set of Q, i.e. $K = \{x \in B \mid Q(x) \leq R^2\}$, then the second property of classical Lyapunov functions, namely $Q(x) > Q(x_0)$ for all $x \in K \setminus \{x_0\}$, is a consequence as we show in Theorem 2.24, x_0 is asymptotically stable and K is a subset of the basin of attraction $A(x_0)$. We call such a compact sublevel set K a Lyapunov basin.

Definition 2.23 (Lyapunov basin). *Let x_0 be an equilibrium of $\dot{x} = f(x)$, $f \in C^1(\mathbb{R}^n, \mathbb{R}^n)$. Let $B \ni x_0$ be an open set. A function $Q \in C^1(B, \mathbb{R})$ and a compact set $K \subset B$ are called a* Lyapunov function Q *with* Lyapunov basin K *if*

1. $x_0 \in \overset{\circ}{K}$,
2. $Q'(x) < 0$ *holds for all* $x \in K \setminus \{x_0\}$,
3. $K = \{x \in B \mid Q(x) \leq R^2\}$ *for an* $R \in \mathbb{R}^+$, *i.e.* K *is a sublevel set of* Q.

We assume without loss of generality that $Q(x_0) = 0$, this can be achieved by adding a constant. In this situation we define the following sublevel sets of Q for $0 < r \leq R$:

$$\begin{aligned} \tilde{B}_r^Q(x_0) &:= \{x \in B \mid Q(x) < r^2\} \\ \text{and } \tilde{K}_r^Q(x_0) &:= \{x \in B \mid Q(x) \leq r^2\} \end{aligned}$$

Note that the function Q of Definition 2.23 is in particular a Lyapunov function in the sense of Definition 2.21. If the equilibrium x_0 is exponentially asymptotically stable and Q is a Lyapunov function in the sense of Definition 2.21, then there exists a corresponding Lyapunov basin K, cf. Theorem 2.45.

The following well-known theorem provides a sufficient condition for a compact set K to belong to the basin of attraction using a Lyapunov function $Q(x)$ (cf. for instance [35], p. 157): in short, a Lyapunov basin is a subset of the basin of attraction.

Theorem 2.24. *Let x_0 be an equilibrium of $\dot{x} = f(x)$ with $f \in C^1(\mathbb{R}^n, \mathbb{R}^n)$. Let Q be a Lyapunov function with Lyapunov basin K in the sense of Definition 2.23.*

Then x_0 is asymptotically stable, K is positively invariant and

$$K \subset A(x_0)$$

holds. Moreover, $Q(z) > Q(x_0)$ holds for all $z \in B \setminus \{x_0\}$.

PROOF: First, we show that the compact set K is positively invariant: Assuming the opposite, there is an $x \in \partial K$ and a $T > 0$ such that $S_T x \notin K$. By assumption, there is an open neighborhood $U \subset B$ of x such that $Q'(y) < 0$ holds for all $y \in U$ and, without loss of generality, we assume $S_t x \in U$ for all $t \in [0, T]$. Then $R^2 < Q(S_T x) = Q(x) + \int_0^T Q'(S_\tau x)\, d\tau < R^2 + 0$, contradiction. Hence, K is positively invariant.

Now we fix an $x \in K$. For the ω-limit set $\emptyset \neq \omega(x) \subset K$ holds by Lemma 2.14. LaSalle's principle Theorem 2.20 implies $Q'(z) = 0$ for all $z \in \omega(x)$. By Definition 2.23, 2. this property only holds for $x = x_0$, hence $\omega(x) = \{x_0\}$ holds for all $x \in K$.

Next, we show that $Q(z) > Q(x_0)$ holds for all $z \in B \setminus \{x_0\}$: For $z \in B \setminus K$ the statement follows by the Definition 2.23, 3. For $z \in K \setminus \{x_0\}$ we have $\omega(z) = \{x_0\}$ and hence there is a sequence t_k, which can be assumed to be strictly monotonously increasing with $t_1 > 0$ without loss of generality, such that $\lim_{k \to \infty} S_{t_k} z = x_0$. Since Q is continuous, we have $\lim_{k \to \infty} Q(S_{t_k} z) = Q(x_0)$. Moreover, $Q(z) > Q(S_{t_1} z) \geq Q(S_{t_k} z)$ holds for all $k \in \mathbb{N}$, so that $Q(z) > Q(S_{t_1} z) \geq Q(x_0)$.

Now we show the stability of x_0: Assuming the opposite, there is an $\epsilon > 0$, a sequence $x_k \in K$ with $\lim_{k \to \infty} x_k = x_0$ and a sequence $t_k \geq 0$, such that $S_{t_k} x_k \notin B_\epsilon(x_0)$ holds. Since K is positively invariant, we have $S_t x_k \in K$ for all $t \in [0, t_k]$. As $Q(x)$ is not increasing as long as x is in K, we have $Q(x_k) \geq Q(S_{t_k} x_k)$. Since $K \setminus B_\epsilon(x_0)$ is compact, there is a subsequence of $S_{t_k} x_k$ which converges to a $z \in K \setminus B_\epsilon(x_0)$. Thus, for $k \to \infty$ we obtain $Q(x_0) \geq Q(z)$. But for $z \in K \setminus \{x_0\}$ we have just shown $Q(z) > Q(x_0)$: contradiction.

Let $x \in K$. Since x_0 is stable and $\omega(x) = \{x_0\}$ holds, $x \in A(x_0)$, cf. Lemma 2.17. In particular, x_0 is asymptotically stable. □

Later we will have the situation that Q and K are as in Definition 2.23 except for the fact that $Q'(x) < 0$ only holds for all $x \in K \setminus E$ with an exceptional set E. If $E \subset A(x_0)$, then $K \subset A(x_0)$ still holds as we show in Theorem 2.26. Note that we do not claim $E \subset K$.

Definition 2.25 (Non-local Lyapunov function, Lyapunov basin and exceptional set). *Let x_0 be an asymptotically stable equilibrium of $\dot{x} = f(x)$, $f \in C^1(\mathbb{R}^n, \mathbb{R}^n)$.*

Let $B \ni x_0$ be an open set. A function $Q \in C^1(B, \mathbb{R})$, a compact set $K \subset B$ and an open set $E \ni x_0$ are called non-local Lyapunov function Q *with* Lyapunov basin K *and* exceptional set E *if*

1. $x_0 \in \mathring{K}$,
2. $Q'(x) < 0$ *holds for all* $x \in K \setminus E$,
3. $K = \{x \in B \mid Q(x) \leq R\}$ *for an* $R \in \mathbb{R}$, *i.e.* K *is a sublevel set of* Q,
4. $E \subset A(x_0)$.

Theorem 2.26. *Let x_0 be an asymptotically stable equilibrium of $\dot{x} = f(x)$ with $f \in C^1(\mathbb{R}^n, \mathbb{R}^n)$. Let Q be a non-local Lyapunov function with Lyapunov basin K and exceptional set E in the sense of Definition 2.25.*

Then $K \subset A(x_0)$ holds.

PROOF: Let $x \in K$. We want to show that $x \in A(x_0)$. Therefore, we distinguish between the two cases:

1. there is a $T_0 \geq 0$ such that $S_{T_0}x \in E$ and
2. the positive orbit $O^+(x) = \bigcup_{t=0}^{\infty} S_t x \subset \mathbb{R}^n \setminus E$.

In case 1., $S_{T_0}x \in E \subset A(x_0)$ and hence $x \in A(x_0)$. We will now show that case 2. does not occur: Assuming the opposite, there is an x such that case 2. holds, in particular $K \setminus E \neq \varnothing$.

We will show that $\bigcup_{t=0}^{\infty} S_t x \in K \setminus E$. Indeed, assuming the opposite, there is a $T_0 \geq 0$ such that $w := S_{T_0}x \in \partial K \setminus E$ and a $T > 0$ such that $S_T w \notin K$. By assumption, there is an open neighborhood $U \subset B$ of w such that $Q'(y) < 0$ holds for all $y \in U$ and, without loss of generality, we assume $S_t w \in U$ for all $t \in [0, T]$. Then $R < Q(S_T w) = Q(w) + \int_0^T Q'(S_\tau w)\, d\tau < R + 0$, contradiction. Hence, $\bigcup_{t=0}^{\infty} S_t x \in K \setminus E$.

Since $K \setminus E$ is non-empty and compact, $\varnothing \neq \omega(x) \subset K \setminus E$ holds for the ω-limit set by Lemma 2.14. By LaSalle's principle Theorem 2.20 we have $Q'(y) = 0$ for all $y \in \omega(x)$. But by Definition 2.25, 2. $Q'(z) < 0$ holds for all $z \in K \setminus E$, which is a contradiction. Hence, case 2. does not occur and the theorem is proven. □

Theorems 2.24 and 2.26 show that a Lyapunov basin is a subset of the basin of attraction. However, the question arises whether such Lyapunov functions and basins exist, how they can be constructed and whether we can use them to obtain the whole basin of attraction.

First, we focus on the problem of finding a Lyapunov function with negative orbital derivative. We can explicitly construct *local* Lyapunov functions $\mathfrak{q}$ using the linearized system around x_0, i.e. $\mathfrak{q}'(x) < 0$ holds for all $x \in U \setminus \{x_0\}$ where U is a (possibly small) neighborhood of x_0, cf. Section 2.2. On the other hand, we can prove the existence of *global* Lyapunov functions Q, i.e. functions satisfying $Q'(x) < 0$ for all $x \in A(x_0) \setminus \{x_0\}$, cf. Section 2.3. These global Lyapunov functions, however, cannot be calculated explicitly in general. Using the properties of the global Lyapunov functions that we show in this chapter, we will approximate them using radial basis functions in Chapter 4. The approximations then turn out to be Lyapunov functions themselves.

The setting is summarized by the following assumption:

Assumption *We consider the autonomous system of differential equations*

$$\dot{x} = f(x),$$

where $f \in C^\sigma(\mathbb{R}^n, \mathbb{R}^n)$, $\sigma \geq 1$, $n \in \mathbb{N}$. We assume the system to have an exponentially asymptotically stable equilibrium x_0 such that $-\nu < 0$ is the maximal real part of the eigenvalues of $Df(x_0)$.

2.2 Local Lyapunov Functions

There is no general approach to explicitly construct Lyapunov functions for a nonlinear system. However, for *linear* systems we can explicitly calculate Lyapunov functions. Hence, in this section we study Lyapunov functions for linear differential equations of the form $\dot{x} = Ax$. The goal of this section is twofold: on the one hand we thus explicitly construct Lyapunov functions for linear systems. On the other hand, we start from the nonlinear differential equation $\dot{x} = f(x)$ and consider the linearization around the equilibrium, namely

$$\dot{x} = Df(x_0)(x - x_0). \tag{2.4}$$

Hence, we set $A = Df(x_0)$. The Lyapunov functions for the linearized system (2.4) turn out to be Lyapunov functions for the nonlinear system since the behavior of solutions near x_0 of the nonlinear and the linearized system is similar. However, while Lyapunov functions for linear systems are always global Lyapunov functions, i.e. their orbital derivative is negative in $\mathbb{R}^n \setminus \{0\}$, the Lyapunov functions are only local for the nonlinear system, i.e. their orbital derivative is negative in some neighborhood of x_0, which is small in general.

In this section we consider the following Lyapunov functions of the linearized system (2.4), which are quadratic forms in $(x - x_0)$ and hence defined in $\mathbb{R}^n$:

- $\mathfrak{d}$ satisfying $\langle \nabla \mathfrak{d}(x), Df(x_0)(x - x_0)\rangle \leq 2(-\nu + \epsilon)\mathfrak{d}(x)$ with $\epsilon > 0$.
- $\mathfrak{v}$ satisfying $\langle \nabla \mathfrak{v}(x), Df(x_0)(x - x_0)\rangle = -(x - x_0)^T C(x - x_0)$ where C is a positive definite matrix, often $C = I$.

Note that the left-hand sides are the orbital derivatives with respect to the linearized system (2.4). These functions can be calculated explicitly with knowledge of the Jordan normal form (for $\mathfrak{d}$) or as the solution of a matrix equation, i.e. a system of linear equations (for $\mathfrak{v}$).

In the following we write $\mathfrak{q}$ for a local Lyapunov function, where either $\mathfrak{q} = \mathfrak{d}$ or $\mathfrak{q} = \mathfrak{v}$. We show that $\mathfrak{q}$ is a Lyapunov function for the nonlinear system $\dot{x} = f(x)$ and that there are Lyapunov basins $\tilde{K}_r^{\mathfrak{q}}(x_0) := \{x \in \mathbb{R}^n \mid \mathfrak{q}(x) \leq r^2\}$ with $r > 0$ (cf. Definition 2.23), where $\mathfrak{q} = \mathfrak{d}$ or $\mathfrak{q} = \mathfrak{v}$.

Since the proofs of the existence use local properties (linearization) and the Lyapunov basins obtained are small, we call these Lyapunov functions and Lyapunov basins *local*.

2.2.1 The Function $\mathfrak{d}$ (Jordan Normal Form)

In this section we calculate a Lyapunov function for a linear equation using the Jordan normal form. The resulting Lyapunov function does not satisfy an equation for the orbital derivative, but an inequality. In the next lemma we consider a matrix A such that all its eigenvalues have negative real part.

We define a transformation matrix S consisting of the transformation to the real Jordan normal form and a scaling by ϵ in the direction of the generalized eigenvectors. For the nonlinear differential equation $\dot{x} = f(x)$ we later choose $A = Df(x_0)$.

Lemma 2.27. *Let A be an $(n \times n)$ matrix such that $-\nu < 0$ is the maximal real part of the eigenvalues of A. Then for each $\epsilon > 0$ there is an invertible matrix $S = S(\epsilon)$ such that $B := SAS^{-1}$ satisfies*

$$u^T Bu \leq (-\nu + \epsilon)\|u\|^2 \text{ for all } u \in \mathbb{R}^n.$$

PROOF: Denote the real eigenvalues of A by $\alpha_1, \alpha_2, \ldots, \alpha_r \in \mathbb{R}$ and the complex eigenvalues by $\alpha_{r+1}, \overline{\alpha_{r+1}}, \ldots \alpha_{r+c}, \overline{\alpha_{r+c}} \in \mathbb{C}\setminus\mathbb{R}$, and set $\alpha_{r+j} =: \lambda_j + i\mu_j$, where $\lambda_j, \mu_j \in \mathbb{R}$ for $1 \leq j \leq c$. Let $m_1, m_2, \ldots, m_{r+c} \geq 1$ be the lengths of the Jordan blocks, so that $\sum_{i=1}^r m_i + \sum_{j=1}^c 2m_{r+j} = n$. Denote a basis of the corresponding generalized real and complex eigenvectors by

$$w_1^1, \ldots, w_1^{m_1}, \ldots, w_r^1, \ldots, w_r^{m_r}, u_{r+1}^1 + iv_{r+1}^1, \ldots, u_{r+1}^{m_{r+1}} + iv_{r+1}^{m_{r+1}}, u_{r+1}^1 - iv_{r+1}^1$$
$$\ldots, u_{r+1}^{m_{r+1}} - iv_{r+1}^{m_{r+1}}, \ldots, u_{r+c}^1 + iv_{r+c}^1, \ldots, u_{r+c}^{m_{r+c}} + iv_{r+c}^{m_{r+c}}, u_{r+c}^1 - iv_{r+c}^1,$$
$$\ldots, u_{r+c}^{m_{r+c}} - iv_{r+c}^{m_{r+c}}.$$

Let $\tilde{T}$ be the matrix with columns $w_1^1, \ldots, w_1^{m_1}, \ldots, w_r^1, \ldots, w_r^{m_r}, u_{r+1}^1, v_{r+1}^1, \ldots, u_{r+1}^{m_{r+1}}, v_{r+1}^{m_{r+1}}, \ldots, u_{r+c}^1, v_{r+c}^1, \ldots, u_{r+c}^{m_{r+c}}, v_{r+c}^{m_{r+c}}$. By the Jordan normal form theorem $\tilde{T}^{-1}A\tilde{T}$ is the real Jordan normal form of A, i.e.

$$\tilde{T}^{-1}A\tilde{T} = \begin{pmatrix} J_1 & & & & & \\ & \ddots & & & & \\ & & J_r & & & \\ & & & K_{r+1} & & \\ & & & & \ddots & \\ & & & & & K_{r+c} \end{pmatrix} \quad \text{with the Jordan matrices}$$

$$J_i := \begin{pmatrix} \alpha_i & 1 & & 0 \\ & \ddots & \ddots & \\ & & \ddots & 1 \\ & & & \alpha_i \end{pmatrix} \quad \text{and}$$

$$K_{r+j} := \begin{pmatrix} \lambda_j & \mu_j & 1 & & & & & \\ -\mu_j & \lambda_j & 0 & 1 & & & & \\ & & \lambda_j & \mu_j & \ddots & & & \\ & & -\mu_j & \lambda_j & & & & \\ & & & & & & 1 & \\ & & & & \ddots & & 0 & 1 \\ & & & & & & \lambda_j & \mu_j \\ & & & & & & -\mu_j & \lambda_j \end{pmatrix}.$$

Define $\tilde{S}_\epsilon := \begin{pmatrix} S_1 & & & & & \\ & \ddots & & & & \\ & & S_r & & & \\ & & & S_{r+1} & & \\ & & & & \ddots & \\ & & & & & S_{r+c} \end{pmatrix}$, where S_j is given by $S_j := \begin{cases} \mathrm{diag}\left(1, \epsilon, \epsilon^2, \ldots, \epsilon^{m_j-1}\right) & \text{for } j \leq r \\ \mathrm{diag}\left(1, 1, \epsilon, \epsilon, \ldots, \epsilon^{m_j-1}, \epsilon^{m_j-1}\right) & \text{for } j > r \end{cases}$. Setting $S := \left(\tilde{T}\tilde{S}_\epsilon\right)^{-1}$ we get

$$SAS^{-1} = B := \begin{pmatrix} M_1 & & 0 & & & \\ & \ddots & & & 0 & \\ 0 & & M_r & & & \\ & & & Z_{r+1} & & 0 \\ & 0 & & & \ddots & \\ & & & 0 & & Z_{r+c} \end{pmatrix},$$

where

$$M_j := \begin{pmatrix} \alpha_j & \epsilon & & 0 \\ & \ddots & \ddots & \\ & & \ddots & \epsilon \\ & & & \alpha_j \end{pmatrix}$$

$$\text{and } Z_{r+j} := \begin{pmatrix} \lambda_j & \mu_j & \epsilon & & & & & \\ -\mu_j & \lambda_j & 0 & \epsilon & & & & \\ & & \lambda_j & \mu_j & \ddots & & & \\ & & -\mu_j & \lambda_j & & & & \\ & & & & & \epsilon & & \\ & & & & \ddots & 0 & \epsilon \\ & & & & & \lambda_j & \mu_j \\ & & & & & -\mu_j & \lambda_j \end{pmatrix}.$$

Now we prove the inequality: First, let $1 \leq j \leq r$ and $u \in \mathbb{R}^{m_j}$. Then

$$\begin{aligned} & u^T M_j u \\ & = \alpha_j(u_1^2 + u_2^2 + \ldots + u_{m_j}^2) + \epsilon(u_1u_2 + u_2u_3 + \ldots + u_{m_j-1}u_{m_j}) \\ & \leq \alpha_j(u_1^2 + u_2^2 + \ldots + u_{m_j}^2) + \frac{\epsilon}{2}(u_1^2 + u_2^2 + u_2^2 + u_3^2 + \ldots + u_{m_j-1}^2 + u_{m_j}^2) \\ & \leq (\alpha_j + \epsilon)(u_1^2 + u_2^2 + \ldots + u_{m_j}^2). \end{aligned}$$

Now let $1 \leq j \leq c$ and $u \in \mathbb{R}^{2m_{r+j}}$.

$$\begin{aligned}
&u^T Z_{r+j} u \\
&= \lambda_j(u_1^2 + u_2^2 + \ldots + u_{2m_{r+j}}^2) + \mu_j \underbrace{(u_1 u_2 - u_1 u_2 + \ldots)}_{=0} \\
&\quad + \epsilon(u_1 u_3 + u_2 u_4 + u_3 u_5 + \ldots + u_{2m_{r+j}-2} u_{2m_{r+j}}) \\
&\leq \lambda_j(u_1^2 + u_2^2 + \ldots + u_{2m_{r+j}}^2) \\
&\quad + \frac{\epsilon}{2} \underbrace{(u_1^2 + u_3^2 + u_2^2 + u_4^2 + u_3^2 + u_5^2 + \ldots + u_{2m_{r+j}-2}^2 + u_{2m_{r+j}}^2)}_{= u_1^2 + u_2^2 + u_{2m_{r+j}-1}^2 + u_{2m_{r+j}}^2 + 2(u_3^2 + u_4^2 + \ldots + u_{2m_{r+j}-2}^2)} \\
&\leq (\lambda_j + \epsilon)(u_1^2 + u_2^2 + \ldots + u_{2m_{r+j}}^2).
\end{aligned}$$

So we get by definition of B for all $u \in \mathbb{R}^n$

$$\begin{aligned}
u^T B u &\leq \left[\max \left(\max_{1 \leq j \leq r} \alpha_j, \max_{1 \leq i \leq c} \lambda_i \right) + \epsilon \right] (u_1^2 + \ldots + u_n^2) \\
&= (-\nu + \epsilon)\|u\|^2,
\end{aligned}$$

which completes the proof of the lemma. □

In the next lemma we prove that $\mathfrak{d}(x) := \|S(x - x_0)\|^2$ is a Lyapunov function for the nonlinear system in a neighborhood of x_0.

Lemma 2.28. *Consider $\dot{x} = f(x)$ with $f \in C^1(\mathbb{R}^n, \mathbb{R}^n)$. Let x_0 be an equilibrium, such that the maximal real part of all eigenvalues of $Df(x_0)$ is $-\nu < 0$.*

Then for each $\epsilon > 0$ there is an $r > 0$ and an invertible matrix $S = S\left(\frac{\epsilon}{2}\right)$, as defined in Lemma 2.27, such that for

$$\mathfrak{d}(x) := \|S(x - x_0)\|^2 \tag{2.5}$$

the orbital derivative $\mathfrak{d}'(x) = \langle \nabla \mathfrak{d}(x), f(x) \rangle$ satisfies

$$\mathfrak{d}'(x) \leq 2(-\nu + \epsilon)\,\mathfrak{d}(x) \text{ for all } x \in \tilde{K}_r^{\mathfrak{d}}(x_0) = \{x \in \mathbb{R}^n \mid \mathfrak{d}(x) \leq r^2\}. \tag{2.6}$$

Proof: Define $S = S\left(\frac{\epsilon}{2}\right)$ as in Lemma 2.27. Note that $\|S(x - x_0)\|$ is an equivalent norm to $\|x - x_0\|$ since S has full rank. The Taylor expansion gives $f(x) = \underbrace{f(x_0)}_{=0} + Df(x_0)(x - x_0) + \phi(x)$ with a function $\phi \in C^1(\mathbb{R}^n, \mathbb{R})$ such that $\lim_{x \to x_0} \frac{\phi(x)}{\|x - x_0\|} = 0$. Hence, there is an $r > 0$ such that

$$\|\phi(x)\| \leq \frac{\epsilon}{2\|S\|} \|S(x - x_0)\|$$

holds for all $x \in \tilde{K}_r^{\mathfrak{d}}(x_0)$. For $x \in \tilde{K}_r^{\mathfrak{d}}(x_0)$ we thus obtain, cf. (2.5),

$$\begin{aligned}
\mathfrak{d}'(x) &= 2[S(x - x_0)]^T S f(x) \\
&= 2[S(x - x_0)]^T \underbrace{S Df(x_0) S^{-1}}_{= B} [S(x - x_0)] + 2[S(x - x_0)]^T S\phi(x) \\
&\leq 2\left(-\nu + \frac{\epsilon}{2}\right) \|S(x - x_0)\|^2 + \epsilon \|S(x - x_0)\|^2 \text{ by Lemma 2.27} \\
&= 2(-\nu + \epsilon)\,\mathfrak{d}(x).
\end{aligned}$$

This shows (2.6). □

Hence, we have found a pair of Lyapunov function $\mathfrak{d}$ and Lyapunov basin $\tilde{K}_r^{\mathfrak{d}}(x_0)$.

Corollary 2.29 *Let the assumptions of Lemma 2.28 hold and let* $0 < \epsilon < \nu$.

There is an $r > 0$ *such that the function* $\mathfrak{d}$ *defined in Lemma 2.28 is a Lyapunov function with Lyapunov basin* $\tilde{K}_r^{\mathfrak{d}}(x_0)$ *in the sense of Definition 2.23.*

Moreover, for all $x \in \tilde{K}_r^{\mathfrak{d}}(x_0)$ *and for all* $t \geq 0$

$$\mathfrak{d}(S_t x) \leq e^{2(-\nu+\epsilon)t}\,\mathfrak{d}(x) \qquad \textit{holds.}$$

Proof: Let r be as in Lemma 2.28. Note that $\mathfrak{d}(x)$ and $\tilde{K}_r^{\mathfrak{d}}(x_0)$ are a Lyapunov function and Lyapunov basin as in Definition 2.23 with $B = \mathbb{R}^n$. Hence, $\tilde{K}_r^{\mathfrak{d}}(x_0)$ is positively invariant by Theorem 2.24. For all $x \in \tilde{K}_r^{\mathfrak{d}}(x_0) \setminus \{x_0\}$ we have by (2.6)

$$\begin{aligned} \mathfrak{d}'(S_\tau x) &\leq 2(-\nu+\epsilon)\mathfrak{d}(S_\tau x) \\ \frac{d}{d\tau}\ln \mathfrak{d}(S_\tau x) &\leq 2(-\nu+\epsilon) \end{aligned}$$

for all $\tau \geq 0$. By integration with respect to τ from 0 to $t \geq 0$ we thus obtain

$$\begin{aligned} \ln \mathfrak{d}(S_t x) - \ln \mathfrak{d}(x) &\leq 2(-\nu+\epsilon)t \\ \mathfrak{d}(S_t x) &\leq e^{2(-\nu+\epsilon)t}\,\mathfrak{d}(x). \end{aligned}$$

For $x = x_0$ the statement is trivial. □

For an explicit calculation of $\mathfrak{d}$, cf. Example 6.1 in Chapter 6.

2.2.2 The Function $\mathfrak{v}$ (Matrix Equation)

A classical way to calculate a Lyapunov function for a linear system of differential equations is to solve a matrix equation. Consider the system of linear differential equations $\dot{x} = A(x - x_0)$, where A is an $(n \times n)$ matrix such that all eigenvalues have negative real parts. We denote by $\mathfrak{v}$ the function $\mathfrak{v}(x) = (x-x_0)^T B(x-x_0)$ satisfying $\langle \nabla\mathfrak{v}(x), A(x-x_0)\rangle = -(x-x_0)^T C(x-x_0)$. By a classical theorem there is a unique solution B. If C is symmetric (positive definite), then so is B. This is shown in Theorem 2.30, cf. e.g. [33] or [57], p. 168.

For the meaning of (2.7), note that setting $\mathfrak{v}(x) := (x - x_0)^T B(x - x_0)$ the expression $\langle \nabla\mathfrak{v}(x), A(x-x_0)\rangle$ is the orbital derivative of $\mathfrak{v}(x)$ with respect to the linear system $\dot{x} = A(x - x_0)$. We have $\langle \nabla\mathfrak{v}(x), A(x - x_0)\rangle = (A(x - x_0))^T B(x - x_0) + (x - x_0)^T BA(x - x_0) = (x - x_0)^T (A^T B + BA)(x - x_0) = -(x - x_0)^T C(x - x_0)$. If C is positive definite, then $\langle \nabla\mathfrak{v}(x), A(x - x_0)\rangle \leq 0$ holds for all $x \in \mathbb{R}^n$.

Theorem 2.30. *Let A be an $(n \times n)$ matrix such that $-\nu < 0$ is the maximal real part of all eigenvalues of A. Consider the linear differential equation*

$$\dot{x} = A(x - x_0)$$

Then for each matrix C there exists one and only one solution B of (2.7):

$$A^T B + BA = -C. \tag{2.7}$$

If C is positive definite (symmetric), then so is B.

PROOF: [[33] or [57], p. 168] $B \to A^T B + BA$ is a linear mapping from $\mathbb{R}^{n\times n}$ to $\mathbb{R}^{n\times n}$. For linear mappings between finite-dimensional vector spaces of same dimension, injectivity and surjectivity are equivalent. Thus, we only show that the mapping is surjective.

We consider (2.7) with given matrices A and C, and show that there is a solution B. Since the maximal real part of all eigenvalues of A is $-\nu < 0$, there is a constant $c > 0$ such that $\|e^{tA^T} C e^{tA}\| \le c e^{-\nu t}$ holds for all $t \ge 0$. Thus, $B := \int_0^\infty e^{tA^T} C e^{tA}\, dt$ exists. We show that B is a solution of (2.7):

$$\begin{aligned}
A^T B + BA &= \int_0^\infty \left[A^T e^{tA^T} C e^{tA} + e^{tA^T} C e^{tA} A\right] dt \\
&= \int_0^\infty \frac{d\left[e^{tA^T} C e^{tA}\right]}{dt}\, dt \\
&= \lim_{t\to\infty} e^{tA^T} C e^{tA} - C \\
&= -C.
\end{aligned}$$

By definition of B it is clear that if C is symmetric (positive definite), then so is B. □

Remark 2.31 *For given A and C, the matrix B can be calculated by solving the system of n^2 linear equations (2.7). Note that if C is symmetric, then the number of equations can be reduced to $n + (n-1) + \ldots + 1 = \frac{(n+1)n}{2}$.*

For the nonlinear system $\dot{x} = f(x)$, the function $\mathfrak{v}$ corresponding to the linearized system turns out to be a Lyapunov function for a small neighborhood of the equilibrium x_0 and satisfies $\mathfrak{v}'(x) = -(x - x_0)^T C(x - x_0) + \tilde{\epsilon}(x)$ with $\lim_{x\to x_0} \frac{\tilde{\epsilon}(x)}{\|x - x_0\|^2} = 0$. The proof uses the Taylor expansion around x_0. In Theorem 2.32 and its Corollary 2.33 we prove, more generally, the existence of a local Lyapunov function with exponential decay.

Theorem 2.32. *Consider $\dot{x} = f(x)$ with $f \in C^1(\mathbb{R}^n, \mathbb{R}^n)$. Let x_0 be an equilibrium, such that the maximal real part of all eigenvalues of $Df(x_0)$ is $-\nu < 0$.*

Then for all $0 \le \tilde{\nu} < 2\nu$ and all symmetric and positive definite $(n \times n)$-matrices C there is a symmetric and positive definite $(n \times n)$-matrix B and *a function* $\tilde{\epsilon} \in C^1(\mathbb{R}^n, \mathbb{R})$ *with* $\lim_{x \to x_0} \frac{\tilde{\epsilon}(x)}{\|x-x_0\|^2} = 0$ *such that for* $\mathfrak{v}(x) := (x-x_0)^T B(x-x_0)$ *we have*

$$\mathfrak{v}'(x) = -\tilde{\nu}\,\mathfrak{v}(x) - (x-x_0)^T C(x-x_0) + \tilde{\epsilon}(x). \tag{2.8}$$

In particular, there is an $r > 0$ *such that*

$$\mathfrak{v}'(x) \leq -\tilde{\nu}\,\mathfrak{v}(x) \tag{2.9}$$

holds for all $x \in \tilde{K}_r^{\mathfrak{v}}(x_0)$.

The matrix B *is the unique solution of the matrix equation*

$$Df(x_0)^T B + BDf(x_0) = -C - \tilde{\nu}B. \tag{2.10}$$

PROOF: Set $A := Df(x_0) + \frac{1}{2}\tilde{\nu}I$. We show that A has only eigenvalues $\tilde{\lambda}$ with $\operatorname{Re}\tilde{\lambda} < 0$. Indeed, let $\tilde{\lambda}$ be an eigenvalue of A. Then $\lambda := \tilde{\lambda} - \frac{1}{2}\tilde{\nu}$ is an eigenvalue of $Df(x_0)$. But for these eigenvalues $\operatorname{Re}\tilde{\lambda} - \frac{1}{2}\tilde{\nu} \leq -\nu$ holds by assumption, hence $\operatorname{Re}\tilde{\lambda} \leq \frac{\tilde{\nu}-2\nu}{2} < 0$. By Theorem 2.30, the equation $A^T B + BA = -C$, which is equivalent to (2.10), has a unique solution B, which is symmetric and positive definite.

The Taylor expansion for the C^1-function f reads $f(x) = f(x_0) + Df(x_0)(x-x_0) + \phi(x)$, where $\phi \in C^1(\mathbb{R}^n, \mathbb{R})$ fulfills $\lim_{x\to x_0} \frac{\phi(x)}{\|x-x_0\|} = 0$. Since $f(x_0) = 0$ we have

$$\begin{aligned}
\mathfrak{v}'(x) &= f(x)^T B(x-x_0) + (x-x_0)^T Bf(x) \\
&= [Df(x_0)(x-x_0) + \phi(x)]^T B(x-x_0) \\
&\quad + (x-x_0)^T B[Df(x_0)(x-x_0) + \phi(x)] \\
&= (x-x_0)^T [Df(x_0)^T B + BDf(x_0)](x-x_0) \\
&\quad + \phi(x)^T B(x-x_0) + (x-x_0)^T B\phi(x) \\
&= -\tilde{\nu}(x-x_0)^T B(x-x_0) - (x-x_0)^T C(x-x_0) + \tilde{\epsilon}(x)
\end{aligned}$$

by (2.10), where $\tilde{\epsilon}(x) := \phi(x)^T B(x-x_0) + (x-x_0)^T B\phi(x)$. Because of the properties of ϕ, $\lim_{x\to x_0} \frac{\tilde{\epsilon}(x)}{\|x-x_0\|^2} = 0$ holds; this shows (2.8).

Thus, for all $\epsilon > 0$ there is an $r > 0$, such that $\|\tilde{\epsilon}(x)\| \leq \epsilon(x-x_0)^T C(x-x_0)$ holds for all $x \in \tilde{K}_r^{\mathfrak{v}}(x_0)$ since $\|x-x_0\|$, $[(x-x_0)^T B(x-x_0)]^{\frac{1}{2}}$ and $[(x-x_0)^T C(x-x_0)]^{\frac{1}{2}}$ are equivalent norms. Choose $\epsilon \leq 1$ and (2.9) follows from (2.8). □

In $\tilde{K}_r^{\mathfrak{v}}(x_0)$ we have an exponential decay of $\mathfrak{v}$, cf. (2.9). Hence, in Corollary 2.33 we obtain the well-known result that the exponential asymptotic stability with respect to the linearized equation implies the exponential asymptotic stability with respect to the nonlinear equation.

Corollary 2.33 *Under the assumptions of Theorem 2.32, for all* $0 \leq \tilde{\nu} < 2\nu$ *there is an* $r > 0$ *such that the function* $\mathfrak{v}$ *is a Lyapunov function with Lyapunov basin* $\tilde{K}_r^{\mathfrak{v}}(x_0)$ *in the sense of Definition 2.23.*

Moreover, for all $x \in \tilde{K}_r^{\mathfrak{v}}(x_0)$ *and for all* $t \geq 0$

$$\mathfrak{v}(S_t x) \leq e^{-\tilde{\nu} t}\mathfrak{v}(x) \qquad \textit{holds.}$$

PROOF: Let r be as in Theorem 2.32. For all $x \in \tilde{K}_r^{\mathfrak{v}}(x_0)$ we have $\mathfrak{v}'(x) \leq -\tilde{\nu}\,\mathfrak{v}(x)$ by (2.9) which implies $\mathfrak{v}(S_t x) \leq e^{-\tilde{\nu} t}\,\mathfrak{v}(x)$ for all $t \geq 0$ as in the proof of Corollary 2.29. $\mathfrak{v}$ is a Lyapunov function with Lyapunov basin $\tilde{K}_r^{\mathfrak{v}}(x_0)$ by (2.9). □

Remark 2.34 *In the following we denote by* $\mathfrak{v}$ *the function as in Corollary 2.33 with the constant* $\tilde{\nu} = 0$ *and* $C = I$, *i.e.* $\mathfrak{v}(x) = (x - x_0)^T B(x - x_0)$ *and*

$$\langle \nabla \mathfrak{v}(x), Df(x_0)(x - x_0)\rangle = -\|x - x_0\|^2.$$

B is then the solution of

$$Df(x_0)^T B + BDf(x_0) = -I.$$

For the explicit calculation of the function $\mathfrak{v}$, cf. Example 6.1 in Chapter 6 or the following example.

2.2.3 Summary and Example

The explicit construction of Lyapunov functions is not possible in general. For linear systems, as explained above, one can explicitly construct Lyapunov functions. Thus, for a nonlinear system, one can use a Lyapunov function for the linearized system, which we call *local Lyapunov function* – this function has negative orbital derivative in a certain neighborhood of x_0. In particular, one can find a corresponding Lyapunov basin, which we call *local Lyapunov basin*. Such a local Lyapunov basin, however, may be very small.

We will illustrate all important steps of our method by an example throughout the book. Namely, we consider the two-dimensional system

$$\begin{cases} \dot{x} = x\left(-1 + 4x^2 + \frac{1}{4}y^2\right) + \frac{1}{8}y^3 \\ \dot{y} = y\left(-1 + \frac{5}{2}x^2 + \frac{3}{8}y^2\right) - 6x^3. \end{cases} \tag{2.11}$$

The system (2.11) has an asymptotically stable equilibrium at $x_0 = (0, 0)$, since the Jacobian at $(0, 0)$ is given by $Df(0, 0) = \begin{pmatrix} -1 & 0 \\ 0 & -1 \end{pmatrix}$ which has the double negative eigenvalue -1.

For (2.11) we calculate the local Lyapunov function $\mathfrak{v}(x, y) = \frac{1}{2}(x^2 + y^2)$, cf. Remark 2.34: with $B = \begin{pmatrix} 1/2 & 0 \\ 0 & 1/2 \end{pmatrix}$ we have $Df(0,0)^T B + BDf(0,0) = -I$, where f denotes the right-hand side of (2.11). We obtain the local Lyapunov basin $\tilde{K} = \{(x, y) \in \mathbb{R}^2 \mid \mathfrak{v}(x, y) \leq 0.09\}$. Figure 2.1 shows the sign of the orbital derivative $\mathfrak{v}'(x, y)$ and the sublevel set $\tilde{K}$ which is bounded by a circle; since the local Lyapunov function is a quadratic form, the level sets are ellipses in general.

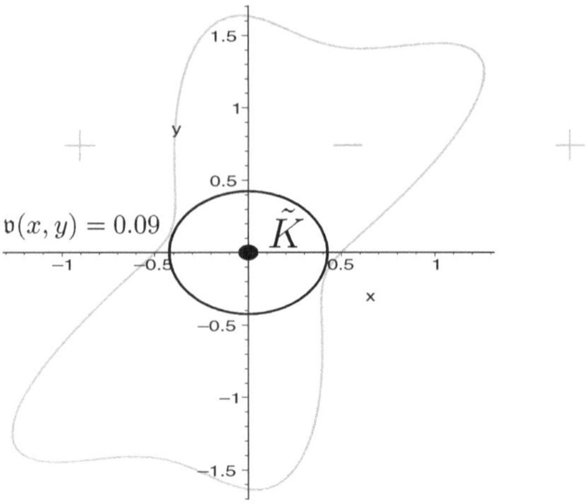

Fig. 2.1. The sign of $\mathfrak{v}'(x, y)$ (grey) and the level set $\mathfrak{v}(x, y) = 0.09$ (black), the boundary of the local Lyapunov basin $\tilde{K}$. The example considered is (2.11).

2.3 Global Lyapunov Functions

Now we turn to the existence of global Lyapunov functions, i.e. functions Q, such that $Q'(x) < 0$ holds for all $x \in A(x_0) \setminus \{x_0\}$.

Converse theorems which guarantee the existence of a Lyapunov function under certain conditions have been given by many authors, for an overview cf. [35] or [58]. The first main converse theorem for asymptotic stability was given by Massera. In [50] he showed that for $f \in C^1$ and $A(x_0) = \mathbb{R}^n$ there exists a C^1 Lyapunov function Q. Later he showed the same result for $f \in C^0$ and $Q \in C^\infty$. The Lyapunov function Q is given by an improper integral over the solution of the differential equation. Barbašin showed $Q \in C^\sigma$ if $f \in C^\sigma$ with dynamical system's methods, cf. [34].

Improvements have been made in several directions, one of which is to prove the existence of smooth Lyapunov functions under weak smoothness assumptions on f. By smoothing techniques of [65], the existence of C^∞-Lyapunov functions was shown, if f is continuous and Lipschitz [47], and even for discontinuous f, cf. [15] and [58]. These results also cover control systems.

A different direction, in which we are particularly interested in, deals with the existence of smooth Lyapunov functions with certain, known values of their orbital derivatives. Here Bhatia [8], cf. also [9] Theorem V. 2.9, showed that there exists a Lyapunov function satisfying $L(x_0) = 0$ and $L(x) > 0$ for all $x \in A(x_0)$ as well as

$$L(S_t x) = e^{-t} L(x) \tag{2.12}$$

for all $x \in A(x_0)$ and $t \in \mathbb{R}$. He assumes that the flow S_t, mapping the initial value $x(0)$ to the solution $x(t)$ at time t, is a continuous mapping for all $t \in \mathbb{R}$ and that x_0 is asymptotically stable. This result is important for our purpose since (2.12) is equivalent to $L'(x) = -L(x)$ if L is smooth enough. The idea of the proof is repeated in this book for the construction of the function T which satisfies $T(x) = \ln L(x)$ and thus $T'(x) = -1$. Since $L(x_0) = 0$, T is not defined in x_0. Note that Bhatia only assumed x_0 to be asymptotically stable whereas x_0 is exponentially asymptotically stable in our case which simplifies the proof considerably.

A different approach is needed to prove the existence of a Lyapunov function V which is also defined in x_0. V satisfies $V'(x) = -p(x)$, where p is a given function with $p(x_0) = 0$. The function V is then given by $V(x) = \int_0^\infty p(S_t x)\, dt$. If p satisfies certain conditions and $f \in C^\sigma(\mathbb{R}^n, \mathbb{R}^n)$, then $V \in C^\sigma(A(x_0), \mathbb{R})$. The idea for this proof goes back to Zubov [70], cf. also Hahn [34], p. 69.

Zubov, moreover, proved that a related Lyapunov function, the solution of a certain partial differential equation, exactly determines the boundary of the basin of attraction. Although the partial differential equation cannot be solved in general, there are some examples where solutions are available, and Zubov also provided a method to approximate the basin of attraction if f is analytic. Generalizations to time-periodic systems and periodic orbits of autonomous systems were given by Aulbach, cf. [3] and [4]. Furthermore, the method was extended to control systems in [13]. For a description of Zubov's method and a review of the determination of the basin of attraction also consider the Russian literature cited in [45].

A first idea for the existence of a Lyapunov function is to use the flow S_T with a fixed $T > 0$ to define a function $Q(x) := \mathfrak{q}(S_T x)$, where $\mathfrak{q} = \mathfrak{d}$ or $\mathfrak{q} = \mathfrak{v}$ denotes one of the local Lyapunov functions with local Lyapunov basin $\tilde{K}_r^{\mathfrak{q}}(x_0)$. The function Q has negative orbital derivatives on $S_{-T}\tilde{K}_r^{\mathfrak{q}}(x_0)$ and thus is not a Lyapunov function on the whole basin of attraction $A(x_0)$ since T is finite. We assume that f is bounded so that S_{-T} is defined on $A(x_0)$, cf. Remark 2.5.

Theorem 2.35. *Consider $\dot{x} = f(x)$ with $f \in C^\sigma(\mathbb{R}^n, \mathbb{R}^n)$, $\sigma \geq 1$ and let x_0 be an equilibrium. Let f be bounded and let the maximal real part of all eigenvalues of $Df(x_0)$ be $-\nu < 0$. Moreover, let $K \subset A(x_0)$ be a compact set.*

Then there is a function $Q \in C^\sigma(A(x_0), \mathbb{R}_0^+)$, such that $Q'(x) < 0$ holds for all $x \in K \setminus \{x_0\}$.

Proof: Since x_0 attracts K uniformly, cf. Proposition 2.11, there is a $T > 0$, such that $S_T K \subset \tilde{K}$ holds with a local Lyapunov basin $\tilde{K} = \tilde{K}_r^{\mathfrak{q}}(x_0)$. With this T we set $Q(x) := \mathfrak{q}(S_T x)$ where $\mathfrak{q}$ is the corresponding local Lyapunov function. We have

$$\begin{aligned}
Q'(x) &= \frac{d}{dt} Q(S_t x)\Big|_{t=0} \\
&= \frac{d}{dt} \mathfrak{q}(S_{T+t} x)\Big|_{t=0} \\
&= \langle \nabla \mathfrak{q}(S_{T+t} x), f(S_{T+t} x)\rangle\Big|_{t=0} \\
&= \langle \nabla \mathfrak{q}(S_T x), f(S_T x)\rangle \\
&= \mathfrak{q}'(S_T x) \\
&< 0
\end{aligned}$$

for $x \in K \setminus \{x_0\}$. This proves the theorem. □

This Lyapunov function has two disadvantages: on the one hand, $Q'(x)$ is not negative in the whole basin of attraction, but only in a compact subset. On the other hand, Q is only as smooth as the flow. In both ways there are better Lyapunov functions: the existence of C^∞-Lyapunov functions for the whole basin of attraction has been shown, if f is continuous and Lipschitz, cf. [47].

A disadvantage of this function and the function of Theorem 2.35 for our purpose is that their orbital derivative satisfies no equation of the form $Q'(x) = -p(x)$ with a known function $p(x)$. This, however, will be important for its approximation and the method of this book. Therefore, we will focus on the two classes of functions T and V with known orbital derivatives.

The natural choice is the class of functions T fulfilling $T'(x) = -\bar{c}$, where $\bar{c}$ is a positive constant. Since the approximation error can be estimated by $|T'(x) - t'(x)| \leq \iota$, we have $t'(x) \leq T'(x) + \iota = -\bar{c} + \iota < 0$ if $\iota < \bar{c}$. However, T is not defined in x_0.

In order to obtain a function which is defined and smooth in $A(x_0)$, we consider the class of functions V satisfying $V'(x) = -p(x)$, where p is a function with several properties, in particular $p(x_0) = 0$. As the function of Theorem 2.35, the functions T and V are only as smooth as the flow. T and V are defined on $A(x_0) \setminus \{x_0\}$, $A(x_0)$, respectively. We will discuss the function T in Section 2.3.1 and the function V in Section 2.3.3.

2.3.1 The Lyapunov Function T with Constant Orbital Derivative

In this section we study the function T which satisfies $T'(x) = -\bar{c}$, where $\bar{c} > 0$. This function is only defined in $A(x_0) \setminus \{x_0\}$ and fulfills $\lim_{x \to x_0} T(x) = -\infty$. Later we define the function L by $L(x) := \exp[T(x)]$ which satisfies $L'(x) = -\bar{c}\, L(x)$. L, however, can be defined at x_0 by $L(x_0) = 0$; the smoothness in x_0 depends on the largest real part $-\nu$ of the eigenvalues of $Df(x_0)$, and L is at least continuous in x_0.

Note that T is the solution of the linear first-order partial differential equation $T'(x) = \sum_{k=1}^n f_k(x) \frac{\partial}{\partial x_k} T(x) = -\bar{c}$. The appropriate problem for this partial differential equation is a Cauchy-problem with an $(n-1)$-dimensional non-characteristic datum manifold Ω and given values H on Ω. The solution of this non-characteristic Cauchy problem is obtained by the characteristic ordinary differential equation: $\dot{x} = f(x)$. Its solutions define a $(n-1)$-parameter family of characteristic curves, parameterized by Ω, which are the orbits of the ordinary differential equation. On Ω the value of T is given; along the characteristic curves the variation is given by $\frac{dT}{dt}(S_t x) = -\bar{c}$. Note that we have a singular point at x_0.

First, we define a *non-characteristic hypersurface* which is the above-mentioned non-characteristic datum manifold in our setting and then we show how to obtain a non-characteristic hypersurface by a Lyapunov function. In Theorem 2.38 we prove the existence of T.

Definition 2.36 (Non-characteristic hypersurface). *Consider $\dot{x} = f(x)$, where $f \in C^\sigma(\mathbb{R}^n, \mathbb{R}^n)$, $\sigma \geq 1$. Let $h \in C^\sigma(\mathbb{R}^n, \mathbb{R})$. The set $\Omega \subset \mathbb{R}^n$ is called a* non-characteristic hypersurface *if*

1. *Ω is compact,*
2. *$h(x) = 0$ if and only if $x \in \Omega$,*
3. *$h'(x) < 0$ holds for all $x \in \Omega$, and*
4. *for each $x \in A(x_0) \setminus \{x_0\}$ there is a time $\theta(x) \in \mathbb{R}$ such that $S_{\theta(x)} x \in \Omega$.*

An example for a non-characteristic hypersurface is the level set of a Lyapunov function within its basin of attraction, e.g. one of the local Lyapunov functions $\mathfrak{d}$ or $\mathfrak{v}$.

Lemma 2.37 (Level sets define a non-characteristic hypersurface). *Let $Q \in C^1(\mathbb{R}^n, \mathbb{R}^n)$ be a Lyapunov function with Lyapunov basin $\tilde{K}_R^Q(x_0) = \{x \in B \mid Q(x) \leq R^2\}$, cf. Definition 2.23, and let $Q(x_0) = 0$ (this can be achieved by adding a constant).*

Then each set $\Omega_r := \{x \in B \mid Q(x) = r^2\}$ with $0 < r \leq R$ is a non-characteristic hypersurface.

PROOF: Set $h(x) := Q(x) - r^2$ on B and extend it smoothly with strictly positive values outside. Then $h(x) = 0$ holds if and only if $x \in \Omega_r$. The set Ω_r is compact, since it is a closed subset of the compact set $\tilde{K}_R^Q(x_0)$. $h'(x) = Q'(x) < 0$ holds for all $x \in \Omega$ by definition of a Lyapunov function.

Now let $x \in A(x_0) \setminus \{x_0\}$. Since $x \in A(x_0)$ there is a time $T \geq 0$ such that $S_T x = z \in \tilde{K}_r^Q(x_0) \setminus \{x_0\}$. Hence, we have $Q(x_0) < Q(z) \leq r^2$. There is a constant $c > 0$ such that we have $Q'(y) \leq -c < 0$ for all $y \in \tilde{K}_R^Q(x_0) \setminus \tilde{B}^Q_{\sqrt{Q(z)}}(x_0) = \{y \in B \mid Q(z) \leq Q(y) \leq R^2\}$, which is a compact set. Thus, for all $t \leq 0$ such that $S_\tau z \in \tilde{K}_R^Q(x_0)$ holds for all $\tau \in [t, 0]$ we have $R^2 \geq Q(S_t z) = Q(z) + \int_0^t Q'(S_\tau z)\, d\tau \geq Q(z) + |t|c$. Hence, by the mean value theorem there is a $t \leq 0$ such that $Q(S_t z) = r^2$ and thus $Q(S_{t+T} x) = r^2$. □

Now we prove the existence theorem of the function T. The proof follows the ideas of [8] or [9], Theorem V. 2.9.

Theorem 2.38 (Existence of T). *Let $\dot{x} = f(x)$, $f \in C^\sigma(\mathbb{R}^n, \mathbb{R}^n)$, $\sigma \geq 1$. Let x_0 be an equilibrium such that $-\nu < 0$ is the maximal real part of all eigenvalues of $Df(x_0)$.*

Let Ω be a non-characteristic hypersurface. Then there is a function $\theta \in C^\sigma(A(x_0) \setminus \{x_0\}, \mathbb{R})$ satisfying

$$S_t x \in \Omega \Leftrightarrow t = \theta(x). \tag{2.13}$$

Furthermore, $\theta'(x) = -1$ and $\lim_{x \to x_0} \theta(x) = -\infty$.

For all $\bar{c} \in \mathbb{R}^+$ and all functions $H \in C^\sigma(\Omega, \mathbb{R})$ there is a function $T \in C^\sigma(A(x_0) \setminus \{x_0\}, \mathbb{R})$ satisfying

$$\begin{aligned} T'(x) &= -\bar{c} \ \text{ for all } x \in A(x_0) \setminus \{x_0\} \text{ and} \\ T(x) &= H(x) \text{ for all } x \in \Omega. \end{aligned}$$

Moreover, $\lim_{x \to x_0} T(x) = -\infty$.

PROOF: We first show that the function θ, implicitly defined by

$$h(S_{\theta(x)} x) = 0,$$

is well-defined on $A(x_0) \setminus \{x_0\}$ and C^σ.

By definition of a non-characteristic hypersurface, there is a $\theta \in \mathbb{R}$ such that $S_\theta x \in \Omega$ and thus $h(S_\theta x) = 0$ holds. We show that θ is unique and, hence, $\theta(x)$ is well-defined: Indeed, assume that $h(S_\theta x) = h(S_{\theta+\tau^*} x) = 0$ for $\theta \in \mathbb{R}$ and $\tau^* > 0$. Since $h'(S_\theta x) < 0$, $h(S_{\theta+t} x) < 0$ for $t \in (0, \tau)$ with some $\tau > 0$. Hence, there is a minimal $\tau^* > 0$ as above such that $h(S_{\theta+\tau^*} x) = 0$. Then $h(S_{\theta+t} x) < 0$ for $t \in (0, \tau^*)$ in contradiction to $h'(S_{\theta+\tau^*} x) < 0$. Hence, the function $\theta(x)$ is well-defined.

We show that $\theta \in C^\sigma(A(x_0) \setminus \{x_0\}, \mathbb{R})$ using the implicit function theorem. θ is the solution t of

$$F(x, t) = h(S_t x) = 0.$$

Let (x, t) be a point satisfying $S_t x \in \Omega$, i.e. $h(S_t x) = 0$. Then

$$\frac{\partial F}{\partial t}(x, t) = h'(S_t x) < 0.$$

Since $S_t x$ is a C^σ function with respect to x and t, cf. [36], $\theta \in C^\sigma$ follows with the implicit function theorem, cf. e.g. [1].

By definition $\theta(S_t x) = \theta(x) - t$. Thus,

$$\theta'(x) = \left. \frac{d}{dt} \theta(S_t x) \right|_{t=0} = -1.$$

Also, $\lim_{x \to x_0} \theta(x) = -\infty$. Indeed, assuming the opposite, there is a sequence $x_k \neq x_0$ with $\lim_{k \to \infty} x_k = x_0$ and $\theta(x_k) \geq -T_0 \in \mathbb{R}^-$. Ω is a compact set by definition and $x_0 \notin \Omega$. Thus, there is an $\epsilon > 0$ with $B_\epsilon(x_0) \cap \Omega = \varnothing$.

Since x_0 is stable, there is a positively invariant neighborhood U of x_0 with $U \subset B_\epsilon(x_0)$ by Lemma 2.16. By continuity of S_{-T_0} there is a $\delta > 0$ such that $S_{-T_0} B_\delta(x_0) \subset U$ since x_0 is an equilibrium and thus a fixed point of S_{-T_0}. Since U is positively invariant, $S_t B_\delta(x_0) \subset U$ holds for all $t \geq -T_0$. Choose $k \in \mathbb{N}$ so large that $x_k \in B_\delta(x_0)$. Then $S_t x_k \notin \Omega$ for $t \in [-T_0, \infty)$ and hence $\theta(x_k) < -T_0$ in contradiction to the assumption.

Define

$$T(x) := \bar{c}\,\theta(x) + H(S_{\theta(x)}x). \tag{2.14}$$

The function is C^σ, satisfies $T'(x) = -\bar{c}$ and $T(x) = H(x)$ for $x \in \Omega$. $\lim_{x \to x_0} T(x) = -\infty$ holds since $\bar{c} > 0$ and the term $H(S_{\theta(x)}x)$ is bounded by $\max_{\xi \in \Omega} |H(\xi)|$. □

We have the following corollary showing that θ and T correspond to the time which a solution takes from one point to the other. Level sets of θ or T thus provide information about the time which solutions take from one level set to another.

Corollary 2.39 *Let the assumptions of Theorem 2.38 hold. If $x, y \in A(x_0) \setminus \{x_0\}$ lie on the same trajectory, then there is a $\tau \in \mathbb{R}$ such that $y = S_\tau x$.*

With the functions θ, T of Theorem 2.38 we have

$$\tau = \theta(x) - \theta(y) = \frac{T(x) - T(y)}{\bar{c}}$$

Proof: We have shown $\theta(S_\tau x) = \theta(x) - \tau$. Also, cf. (2.14), we have

$$T(x) - T(y) = \bar{c}\,[\theta(x) - \theta(y)] + \underbrace{H(S_{\theta(x)}x) - H(S_{\theta(y)}y)}_{=0}$$

since x and y lie on the same trajectory. This shows the corollary. □

Corollary 2.40 *Under the assumptions of Theorem 2.38 there is a function $L \in C^0(A(x_0), \mathbb{R}) \cap C^\sigma(A(x_0) \setminus \{x_0\}, \mathbb{R})$ satisfying*

$$L'(x) = -\bar{c}\,L(x) \qquad \text{for all } x \in A(x_0).$$

Moreover, $L(x) \geq 0$ holds for all $x \in A(x_0)$; $L(x) = 0$ holds if and only if $x = x_0$.

Proof: Set $L(x) := e^{T(x)}$ for $x \in A(x_0) \setminus \{x_0\}$ and $L(x_0) = 0$. L is continuous in x_0 since $\lim_{x \to x_0} T(x) = -\infty$ and thus $\lim_{x \to x_0} L(x) = 0$. □

2.3.2 Level Sets of Lyapunov Functions

The structure of general Lyapunov functions concerning their level sets has been studied by several authors. If one only assumes that the equilibrium is *asymptotically stable*, then Wilson [64] showed that the level sets are homotopically equivalent to a sphere. Thus, the level sets are diffeomorphic to a sphere if $n \neq 4, 5$. However, for any space dimension n, $A(x_0)$ is diffeomorphic to $\mathbb{R}^n$, cf. [64]. The homotopical equivalence is used in Grüne et al. [31] to show that one can transform an asymptotically stable system into an exponentially asymptotically stable system if $n \neq 4, 5$. We, however, assume that the equilibrium is *exponentially* asymptotically stable and thus can prove that the level sets of a smooth Lyapunov function are diffeomorphic to spheres in any dimension.

This result has important implications for our method: given a Lyapunov function, i.e. a function with negative orbital derivative in a certain set $K \setminus \{x_0\}$, one can always find a corresponding Lyapunov basin, i.e. a sublevel set of the Lyapunov function, and thus obtain a subset of the basin of attraction. Moreover, this Lyapunov basin is maximal in the sense that it comes arbitrarily near to the boundary of K, if K is compact.

In this section we consider a general Lyapunov function Q for an exponentially asymptotically stable equilibrium x_0. We show that level sets of Lyapunov functions Q are diffeomorphic to spheres in Corollary 2.43. In Theorem 2.45 we show that given a Lyapunov function, i.e. a function with negative orbital derivative, one can always find a Lyapunov basin and thus apply Theorem 2.24. The proofs of the results in this section use similar techniques as in Theorem 2.38.

Definition 2.41. *Let Ω be a non-characteristic hypersurface. Let θ be the function of Theorem 2.38. Define*

$$\Omega^i := \{x \in A(x_0) \setminus \{x_0\} \mid \theta(x) < 0\} \cup \{x_0\}.$$

If the equilibrium x_0 is exponentially asymptotically stable, then the level sets are diffeomorphic to S^{n-1}. For dimensions $n \neq 4, 5$ this is even true for Lyapunov functions of equilibria which are not exponentially, but only asymptotically stable, cf. [31] which uses [64]. Our proof for an *exponentially* asymptotically stable equilibrium, however, is much simpler.

Proposition 2.42 *Let Ω be a non-characteristic hypersurface for an exponentially asymptotically stable equilibrium x_0. Then Ω is σ-diffeomorphic to S^{n-1}. Moreover, Ω^i is homeomorphic to $B_1(0)$ and $\Omega^i \cup \Omega$ is homeomorphic to $\overline{B_1(0)}$ such that each level set $\{x \in \Omega^i \mid \theta(x) = c\}$ with $c < 0$ is mapped to a sphere of radius e^c.*

PROOF: Let $\mathfrak{d}(x) = \|S(x - x_0)\|^2$, cf. (2.5), be the local Lyapunov function of Corollary 2.29 with a local Lyapunov basin $\tilde{K}_r^{\mathfrak{d}}(x_0)$. The mapping

$$d_1\colon S^{n-1} \to \Omega', \qquad d_1(y) := rS^{-1}y + x_0$$

is a C^∞-diffeomorphism from S^{n-1} to $\Omega' := \partial \tilde{K}_r^{\mathfrak{q}}(x_0)$, since S has full rank. Indeed, if $\|y\| = 1$, then $\mathfrak{q}(d_1(y)) = \|S(rS^{-1}y + x_0 - x_0)\|^2 = r^2$.

Since Ω is a non-characteristic hypersurface, by Theorem 2.38 there is a function $\theta \in C^\sigma(A(x_0) \setminus \{x_0\}, \mathbb{R})$ such that $S_{\theta(x)}x \in \Omega$.

Now set

$$d_2\colon \Omega' \to \Omega, \qquad d_2(x) := S_{\theta(x)}x.$$

Obviously, d_2 is C^σ. Since also Ω' is a non-characteristic hypersurface by Lemma 2.37, there is a function $\tilde{\theta} \in C^\sigma(A(x_0)\setminus\{x_0\}, \mathbb{R})$ such that $S_{\tilde{\theta}(y)}y \in \Omega'$. Define the function $\tilde{d}_2 \in C^\sigma(\Omega, \Omega')$ by $\tilde{d}_2(y) = S_{\tilde{\theta}(y)}y$. We show that $\tilde{d}_2 = d_2^{-1}$ holds. Indeed, we have $\tilde{d}_2(d_2(x)) = S_{\tilde{\theta}(d_2(x))+\theta(x)}x \in \Omega'$. Since there is a unique time $t \in \mathbb{R}$ such that $S_t x \in \Omega'$ and $t = 0$ is such a time, $\tilde{d}_2(d_2(x)) = x$. $d_2(\tilde{d}_2(y)) = y$ is shown similarly. Thus,

$$d = d_2 \circ d_1 \in C^\sigma(S^{n-1}, \Omega) \tag{2.15}$$

is a C^σ-diffeomorphism with inverse $d^{-1} = d_1^{-1} \circ \tilde{d}_2 \in C^\sigma(\Omega, S^{n-1})$.

Now we prove that there is a homeomorphism $\phi\colon B_1(0) \to \Omega^i$. Define

$$\phi(y) := S_{-\ln\|y\|}\, d\left(\frac{y}{\|y\|}\right) \text{ for } y \neq 0 \tag{2.16}$$

and $\phi(0) := x_0$. Here, d denotes the diffeomorphism $d \in C^\sigma(S^{n-1}, \Omega)$ defined in (2.15). If $\|y\| < 1$, then $\phi(y) = S_t x$ with $t > 0$, and $x \in \Omega$ and hence $\phi(y) \in \Omega^i$. For all sequences $y \to 0$ we have $\lim_{y\to 0} \phi(y) = \lim_{t\to\infty} S_t d\left(\frac{y}{\|y\|}\right) = x_0$ because $d\left(\frac{y}{\|y\|}\right) \in \Omega \subset A(x_0)$ is compact. Hence, ϕ is continuous. Now define

$$\tilde{\phi}(x) := e^{\theta(x)} d^{-1}(S_{\theta(x)}x) \text{ for } x \in A(x_0) \setminus \{x_0\} \tag{2.17}$$

and $\tilde{\phi}(x_0) = 0$. Since $\lim_{x\to x_0} e^{\theta(x)} d^{-1}(S_{\theta(x)}x) = 0$ by Theorem 2.38, $\tilde{\phi}$ is continuous. If $x \in \Omega^i$, then $e^{\theta(x)} < 1$ and hence $\tilde{\phi}(x) \in B_1(0)$. Note that the level set $L_c := \{x \in \Omega^i \mid \theta(x) = c\}$ with $c < 0$ is mapped to $\tilde{\phi}(L_c) = e^c S^{n-1}$, i.e. the sphere of radius e^c.

Moreover, $\tilde{\phi}(\phi(y)) = \tilde{\phi}\left(S_{-\ln\|y\|} d\left(\frac{y}{\|y\|}\right)\right) = \|y\| \frac{y}{\|y\|} = y$ for $y \neq 0$ and $\phi(\tilde{\phi}(x)) = \phi\left(e^{\theta(x)} d^{-1}(S_{\theta(x)}x)\right) = S_{-\theta(x)} S_{\theta(x)} x = x$ for $x \neq x_0$.

The homeomorphism ϕ can be extended to a homeomorphism from $\Omega^i \cup \Omega = \overline{\Omega^i}$ to $\overline{B_1(0)}$ by the same definition (2.16). □

Corollary 2.43 *Let x_0 be an equilibrium of $\dot{x} = f(x)$, $f \in C^1(\mathbb{R}^n, \mathbb{R}^n)$. Let $-\nu < 0$ be the maximal real part of all eigenvalues of $Df(x_0)$. Let Q be a Lyapunov function with Lyapunov basin $\tilde{K} = \tilde{K}_R^Q(x_0)$, cf. Definition 2.23, and $Q(x_0) = 0$.*

Then for all $0 < r \leq R$ *there is a* C^σ*-diffeomorphism*

$$d \in C^\sigma(S^{n-1}, \{x \in B \mid Q(x) = r^2\}).$$

PROOF: Define $\Omega := \{x \in B \mid Q(x) = r^2\}$. Ω is a non-characteristic hypersurface by Lemma 2.37. By Proposition 2.42 there exists a C^σ-diffeomorphism d as stated in the corollary. □

For the next proposition we assume that f is bounded in $A(x_0)$. This can be achieved by considering a modified differential equation, cf. Remark 2.5.

Proposition 2.44 *Let the assumptions of Theorem 2.38 hold. Let additionally* $\sup_{x\in A(x_0)} \|f(x)\| < \infty$ *hold.*

Then

$$K_R := \{x \in A(x_0) \setminus \{x_0\} \mid T(x) \leq R\} \cup \{x_0\}$$

is a compact set in $\mathbb{R}^n$ *for all* $R \in \mathbb{R}$.

PROOF: We assume that there is a constant C such that $\|f(x)\| \leq C$ holds for all $x \in A(x_0)$. If for an $R \in \mathbb{R}$ the set K_R is not compact in $\mathbb{R}^n$, then either K_R is unbounded or not closed.

We first show that all sets K_R are bounded: We show that $K_R \subset \overline{B_{S+\frac{R+S_H}{\bar{c}}C}(0)} \cup \overline{\Omega^i}$, where $S := \max_{\xi\in\Omega} \|\xi\|$, $S_H := \max_{\xi\in\Omega} |T(x)| = \max_{\xi\in\Omega} |H(x)|$, for the definition of H cf. Theorem 2.38. Since $\overline{\Omega^i}$ is the image of the compact set $\overline{B_1(0)}$ under the continuous mapping ϕ by Proposition 2.42, $\overline{\Omega^i}$ is bounded.

If $K_R \subset \overline{\Omega^i}$, then K_R is bounded. Now let $x \in K_R \setminus \overline{\Omega^i}$, i.e. $\theta(x) > 0$. Then, cf. (2.14),

$$\begin{aligned}
x - S_{\theta(x)}x &= -\int_0^{\theta(x)} f(S_\tau x)\, d\tau \\
\|x\| &\leq \|S_{\theta(x)}x\| + \theta(x) \sup_{\xi\in A(x_0)} \|f(\xi)\| \\
&\leq S + \frac{T(x) + S_H}{\bar{c}} \cdot C \\
&\leq S + \frac{R + S_H}{\bar{c}} \cdot C.
\end{aligned}$$

Hence, K_R is bounded.

Now assume that K_R is not closed for an $R \in \mathbb{R}$. Then there is a sequence $x_k \in K_R$, with $\lim_{k\to\infty} x_k = x^* \notin K_R$. By definition of K_R, $x^* \notin A(x_0)$.

Since $\overline{\Omega_i} \subset A(x_0)$ and $A(x_0)$ is open, there is an $\epsilon > 0$ such that $(\overline{\Omega^i})_\epsilon := \{x \in \mathbb{R}^n \mid \operatorname{dist}(x, \overline{\Omega^i}) < \epsilon\}$ satisfies $(\overline{\Omega^i})_\epsilon \subset A(x_0)$. Let $R^* := \frac{R+S_H}{\bar{c}}$. By continuity of $S_{R^*}x$ with respect to x there is a $\delta > 0$ such that $\|S_{R^*}x^* - S_{R^*}y\| < \epsilon$ holds for all y with $\|x^* - y\| < \delta$. Moreover, there is a $k \in \mathbb{N}$ such that $y = x_k$ satisfies $\|x^* - x_k\| < \delta$. Then

$$T(S_{R^*}y) - T(y) = \int_0^{R^*} T'(S_t y)\, dt = -\bar{c}\, R^*.$$

Hence, by (2.14) we have

$$\begin{aligned} \bar{c}\,\theta(S_{R^*}y) + H(S_{\theta(y)}y) &= T(S_{R^*}y) \\ &= -\bar{c}R^* + T(y) \\ &\leq -\bar{c}R^* + R, \text{ since } y = x_k \in K_R \\ \theta(S_{R^*}y) &\leq -R^* + \frac{R + S_H}{\bar{c}} = 0. \end{aligned}$$

Hence, $S_{R^*}y \in \overline{\Omega^i}$ and $S_{R^*}x^* \in (\overline{\Omega^i})_\epsilon \subset A(x_0)$ in contradiction to the assumption. □

In Theorem 2.45 we study the following problem: Assume, we have a function $Q \in C^1(\mathbb{R}^n, \mathbb{R})$ which satisfies $Q'(x) < 0$ in a set K. How can we use this Lyapunov function to obtain a Lyapunov basin and thus, using Theorem 2.24, to determine a subset of the basin of attraction? The answer is that we can always find a Lyapunov basin, which is maximal in the following sense:

Theorem 2.45. *Let x_0 be an equilibrium of $\dot{x} = f(x)$, where $f \in C^1(\mathbb{R}^n, \mathbb{R}^n)$, such that the real parts of $Df(x_0)$ are all negative.*

Let $Q \in C^1(\mathbb{R}^n, \mathbb{R})$ be a function with $Q(x_0) = 0$ (this can be achieved by adding a constant to Q). Let K be a compact set with $x_0 \in \overset{\circ}{K}$ and let

$$Q'(x) < 0 \qquad \text{hold for all } x \in K \setminus \{x_0\}. \tag{2.18}$$

Then there is an open neighborhood $B \subset \overset{\circ}{K}$ of x_0 and a $y \in \partial B \cap \partial K$ such that with $R^* := \sqrt{Q(y)} > 0$ all sets

$$K_R := \{x \in B \mid Q(x) \leq R^2\} \qquad \text{with } 0 < R < R^*$$

are Lyapunov basins. In particular, $\overset{\circ}{K}$ is an open neighborhood of K_R.

Proof: Let $\mathfrak{q}$ be a local Lyapunov function and let $\tilde{K}_r^{\mathfrak{q}}(x_0) \subset K$ with $r > 0$ be a local Lyapunov basin. Denote

$$\rho := \sqrt{\min_{x \in \partial \tilde{K}_r^{\mathfrak{q}}(x_0)} Q(x)} > 0. \tag{2.19}$$

The minimum exists since $\partial \tilde{K}_r^{\mathfrak{q}}(x_0)$ is a compact set, and $\rho \neq 0$ by Theorem 2.24.

Now fix $\rho > \epsilon > 0$ and set

$$K^0 := \{x \in \tilde{B}_r^{\mathfrak{q}}(x_0) \mid Q(x) \leq (\rho - \epsilon)^2\}.$$

K^0 is a Lyapunov basin with Lyapunov function Q. To show this, we prove that K^0 is a compact set in $\mathbb{R}^n$: It is bounded since $K^0 \subset \tilde{B}_r^{\mathfrak{q}}(x_0)$. To show that it is closed, assume the opposite, i.e. that there is a sequence $x_k \in K^0$ with $\lim_{k\to\infty} x_k = x \notin K^0$. By definition of K^0, $x \in \partial\tilde{K}_r^{q}(x_0)$. Hence $Q(x) \geq \rho^2$ by (2.19) which is a contradiction to $(\rho-\epsilon)^2 \geq \lim_{k\to\infty} Q(x_k) = Q(x) \geq \rho^2$.

Next, consider $x \in \partial K^0$, i.e. $Q(x) = (\rho-\epsilon)^2 > 0$. We show that there is a (maximal) $t \leq 0$ with $S_t x \in \partial K$: Assuming the opposite, $S_t x \in K$ holds for all $t \leq 0$. Hence, every sequence $t_k \to -\infty$ satisfies $S_{t_k} x \in K$ and thus has a convergent subsequence which we still denote by t_k such that $\lim_{k\to\infty} S_{t_k} x = y \in K$.

By an argument similar to LaSalle's principle (Theorem 2.20) we will show that $Q'(y) = 0$. Assume in contradiction that $Q'(y) < 0$. Then there is a $\tau > 0$ such that $Q(S_\tau y) < Q(y)$. There is a subsequence of the t_k which we still denote by t_k such that $t_{k+1} - t_k < -\tau$ holds for all k. Thus, $Q(S_{t_{k+1}+\tau} x) \geq Q(S_{t_k} x)$. By continuity of Q we conclude $Q(S_\tau y) \geq Q(y)$ in contradiction to the assumption. Hence, $Q'(y) = 0$.

Since the only point $y \in K$ with $Q'(y) = 0$ is x_0, we conclude $y = x_0$. But since $S_t x \in K$ we have $Q(S_t x) \geq Q(x)$ for all $t \leq 0$ and thus $0 = \lim_{k\to\infty} Q(S_{t_k} x) \geq Q(x) > 0$, contradiction. Hence, for all $x \in \partial K^0$ there is a $T^*(x)$ which is the maximal $T^*(x) \leq 0$ with $S_{T^*(x)} x \in \partial K$. In particular, $Q(S_{T^*(x)} x) \geq (\rho-\epsilon)^2$. Note that $T^*(x)$ is not continuous with respect to x in general.

Now we show that there is a $T^* \leq 0$ such that $T^*(x) \in [T^*, 0]$ holds for all $x \in \partial K^0$. We have for all $x \in \partial K^0$

$$\begin{aligned}\int_0^{T^*(x)} Q'(S_\tau x)\, d\tau &= Q(S_{T^*(x)} x) - Q(x) \\ &\leq \max_{x\in K} Q(x) - \min_{x\in K} Q(x) =: M \end{aligned} \tag{2.20}$$

Note that $\bigcup_{\tau=T^*(x)}^{0} S_\tau x \in K \setminus \overset{\circ}{K^0}$. Indeed, for $\tau \leq 0$ we have $Q(S_\tau x) \geq Q(x) = (\rho-\epsilon)^2$. Thus, $m := \min_{y \in K\setminus \overset{\circ}{K^0}} |Q'(y)| > 0$. Now (2.20) implies $|T^*(x)| \cdot m \leq \int_0^{T^*(x)} Q'(S_\tau x)\, d\tau \leq M$ and thus $|T^*(x)| \leq \frac{M}{m} =: -T^*$ which is a constant independent of x.

Now define

$$R^* := \sqrt{\inf_{x\in\partial K^0} Q(S_{T^*(x)} x)} \geq \rho - \epsilon,$$

since $T^*(x) \leq 0$ and $S_t x \in K$ holds for $x \in \partial K^0$ and $t \in [T^*(x), 0]$. Let $x_k \in \partial K^0$ be a minimizing sequence. Since ∂K^0 is compact, we can assume that $\lim_{k\to\infty} x_k = \xi \in \partial K^0$. Then we have $y_k := S_{T^*(x_k)} x_k \in \partial K$ and hence there is a convergent subsequence such that both $\lim_{k\to\infty} y_k = y \in \partial K$, since ∂K is compact, and $\lim_{k\to\infty} T^*(x_k) = T \in [T^*, 0]$ since $T^*(x_k) \in [T^*, 0]$. This $y \in \partial K$ satisfies the statements of the theorem. We have $Q(y) = \lim_{k\to\infty} Q(y_k) = \lim_{k\to\infty} Q(S_{T^*(x_k)} x_k) = (R^*)^2$, $y = \lim_{k\to\infty} S_{T^*(x_k)} x_k = S_T \xi$ and $Q(S_T \xi) = Q(y) = (R^*)^2$.

Now define $t(x) \in [T^*(x), 0]$ for $x \in \partial K^0$ by

$$Q(S_{t(x)}x) = (R^*)^2.$$

The function $t(x)$ is well-defined since for all $x \in \partial K^0$ we have $Q(S_{T^*(x)}x) \geq (R^*)^2$, $Q(x) = (\rho - \epsilon)^2 \leq (R^*)^2$ and $Q'(S_t x) < 0$ for all $t \in [T^*(x), 0]$ since here $S_t x \in K \setminus \{x_0\}$. Moreover, $T^*(x) \leq t(x)$ and $t(x)$ is a continuous function with respect to x by the implicit function theorem since $Q'(x) < 0$ holds for all $x \in K \setminus \{x_0\}$. Note that $t(\xi) = T$ and hence $y \in \Omega$, where

$$\Omega := \{S_{t(x)}x \mid x \in \partial K^0\}.$$

We show that Ω is a non-characteristic hypersurface: Ω is compact since it is the continuous image of the compact set ∂K^0.

Define $h(x) = Q(x) - (R^*)^2$ for $x \in K$ and prolongate it smoothly and strictly positive outside; then $h(x) = 0$ if and only if $x \in \Omega$, and $h'(x) = Q'(x) < 0$ holds for all $x \in \Omega$ since ∂K^0 is a non-characteristic hypersurface by Lemma 2.37. For $\xi \in A(x_0) \setminus \{x_0\}$ there is a time $\tau \in \mathbb{R}$ such that $S_\tau \xi =: x \in \partial K^0$, hence $S_{t(x)+\tau}\xi \in \Omega$.

By Proposition 2.42, the set Ω is σ-diffeomorphic to S^{n-1} and Ω^i is homeomorphic to $B_1(0)$. Set $B := \Omega^i$. B is open and $y \in \Omega = \partial \Omega^i$. Note that by definition of Ω^i, cf. Definition 2.41, $\Omega^i = \{S_t x \mid x \in \partial K^0, t > t(x)\} \cup \{x_0\}$. Since in particular $t(x) \geq T^*(x)$, $B = \Omega^i \subset K$ and hence $B \subset \mathring{K}$.

Now let $0 < R < R^*$. We show that B is an open neighborhood of K_R, therefore, we show that $K_R \subset \overline{\phi(B_{r^*}(0))} \subset B$, where $r^* < 1$ will be defined below and ϕ is the homeomorphism $B_1(0) \to B$, cf. (2.16). Define $\tilde{T}(x)$ by $Q(S_{\tilde{T}(x)}x) = R^2$ for all $x \in B$. Let $x \in K_R$, i.e. $Q(x) \leq R^2$. Then $\theta(x) \leq \tilde{T}(x) \leq 0$ for $x \in K_R$ where $\theta(x)$ is the function of Theorem 2.38 satisfying $S_{\theta(x)}x \in \Omega$. With $C' := \max_{\xi \in K} |Q'(\xi)| > 0$ we have

$$\begin{aligned}(R^*)^2 - R^2 &= Q(S_{\theta(x)}x) - Q(S_{\tilde{T}(x)}x) \\ &= \int_{\tilde{T}(x)}^{\theta(x)} Q'(S_\tau x)\, d\tau \\ &\leq [\tilde{T}(x) - \theta(x)]C' \\ &\leq -\theta(x)C'.\end{aligned}$$

Thus, $\theta(x) \leq -\frac{(R^*)^2 - R^2}{C'}$ and $\phi^{-1}(K_R) \subset \overline{B_{r^*}(0)}$, cf. (2.17), where we set $r^* = \exp\left(-\frac{(R^*)^2 - R^2}{C'}\right) < 1$. □

2.3.3 The Lyapunov Function V Defined in $A(x_0)$

In this section we show the existence of a Lyapunov function V such that $V' = -p(x)$ holds. $p(x)$ is a given function with $p(x) > 0$ for $x \neq x_0$ and $p(x) = O(\|x - x_0\|^\eta)$ for $x \to x_0$ with $\eta > 0$. Among these functions we consider quadratic forms $(x - x_0)^T C(x - x_0)$ with a positive definite matrix

C and in particular $C = I$, i.e. $p(x) = \|x - x_0\|^2$. The resulting function V is defined on $A(x_0)$ and as smooth as the flow or f is, i.e. $V \in C^\sigma(A(x_0), \mathbb{R})$. This section follows [24]. In Proposition 2.48 we will show that the function V with these properties is unique up to a constant.

Theorem 2.46 (Existence of V). *Let x_0 be an equilibrium of $\dot{x} = f(x)$ with $f \in C^\sigma(\mathbb{R}^n, \mathbb{R}^n)$, $\sigma \geq 1$, such that the maximal real part of all eigenvalues of $Df(x_0)$ is $-\nu < 0$. Let $p \in C^\sigma(\mathbb{R}^n, \mathbb{R})$ be a function with the following properties:*

1. *$p(x) > 0$ for $x \neq x_0$,*
2. *$p(x) = O(\|x - x_0\|^\eta)$ with $\eta > 0$ for $x \to x_0$,*
3. *For all $\epsilon > 0$, p has a positive lower bound on $\mathbb{R}^n \setminus B_\epsilon(x_0)$.*

Then there exists a function $V \in C^\sigma(A(x_0), \mathbb{R})$ with $V(x_0) = 0$ such that

$$V'(x) = -p(x)$$

holds for all $x \in A(x_0)$.

If $\sup_{x \in A(x_0)} \|f(x)\| < \infty$, then

$$K_R := \{x \in A(x_0) \mid V(x) \leq R^2\}$$

is a compact set in $\mathbb{R}^n$ for all $R \geq 0$.

PROOF: Define the function V by

$$V(x) := \int_0^\infty p(S_t x)\, dt \geq 0.$$

By the properties of p, $V(x) = 0$, if and only if $x = x_0$. Provided that the integral and its derivative converge uniformly, we have

$$\begin{aligned} V'(x) &= \frac{d}{d\tau} \lim_{T \to \infty} \int_0^T p(S_{t+\tau} x)\, dt \Big|_{\tau=0} \\ &= \lim_{T \to \infty} \frac{d}{d\tau} \int_\tau^{T+\tau} p(S_t x)\, dt \Big|_{\tau=0} \\ &= \lim_{T \to \infty} p(S_T x) - p(x) \\ &= p(x_0) - p(x) \\ &= -p(x). \end{aligned}$$

We show that the integral and all derivatives of maximal order $\sigma \geq 1$ with respect to x converge uniformly. Then the smoothness follows.

By Corollary 2.29 or 2.33 with the constants $0 < \tilde{\nu} := \frac{4}{3}\nu < 2\nu$ and $\epsilon := \frac{\nu}{3}$ there is a positively invariant and bounded neighborhood $\tilde{K} = \tilde{K}_r^{\mathfrak{q}}(x_0)$ of x_0, such that $\mathfrak{q}(S_t x) \leq e^{-\tilde{\nu} t} \mathfrak{q}(x)$ holds for all $x \in \tilde{K}$ and all $t \geq 0$; here again $\mathfrak{q}$ denotes either $\mathfrak{q} = \mathfrak{d}$ or $\mathfrak{q} = \mathfrak{v}$. Moreover, choose $\tilde{K}$ so small that $p(x) \le c^* \|x - x_0\|^\eta$ holds for all $x \in \tilde{K}$. Since $\mathfrak{q}$ is a quadratic form, there are constants $0 < c_1 \le c_2$ such that $c_1 \, \mathfrak{q}(x) \le \|x - x_0\|^2 \le c_2 \, \mathfrak{q}(x)$ holds for all $x \in \mathbb{R}^n$. Hence, for all $x \in \tilde{K}$ we have

$$\begin{aligned} \|S_t x - x_0\|^\eta &\le (c_2 \mathfrak{q}(S_t x))^{\frac{\eta}{2}} \\ &\le \left(c_2 e^{-\tilde{\nu} t} \mathfrak{q}(x)\right)^{\frac{\eta}{2}} \\ &\le \underbrace{c_2^{\frac{\eta}{2}}}_{:= C^*} \mathfrak{q}(x)^{\frac{\eta}{2}} e^{-\tilde{\nu}\frac{\eta}{2} t}. \end{aligned} \tag{2.21}$$

Let $x \in A(x_0)$ and let O be a bounded, open neighborhood of x, such that $\overline{O} \subset A(x_0)$. Since $\overline{O}$ is compact and $\tilde{K}$ positively invariant, Proposition 2.11 implies that there is a $T \in \mathbb{R}_0^+$ such that $S_{T+t}\overline{O} \subset \tilde{K}$ holds for all $t \ge 0$. Thus, we have for all $z \in \overline{O}$

$$\begin{aligned} &\int_0^\infty p(S_t z)\, dt \\ &= \max_{y \in \overline{O}} \int_0^T p(S_t y)\, dt + \max_{y \in \tilde{K}} \int_0^\infty p(S_t y)\, dt \\ &\le \max_{y \in \overline{O}} \int_0^T p(S_t y)\, dt + c^* C^* \int_0^\infty e^{-\tilde{\nu}\frac{\eta}{2} t}\, dt \left(\max_{y \in \tilde{K}} \mathfrak{q}(y)\right)^{\frac{\eta}{2}} \quad \text{by (2.21)} \\ &= \max_{y \in \overline{O}} \int_0^T p(S_t y)\, dt + \frac{2c^* C^*}{\tilde{\nu}\eta} r^\eta \end{aligned}$$

since $\tilde{K} = \{y \in \mathbb{R}^n \mid \mathfrak{q}(y) \le r^2\}$. Hence, the integral converges uniformly on $\overline{O}$. Similarly, $p(S_t x)$ converges to $p(x_0) = 0$ as $t \to \infty$, uniformly for all $x \in \overline{O}$.

Now we show that $V \in C^\sigma(A(x_0), \mathbb{R})$; the proof of this fact needs several preliminary steps. First we prove a formula for the derivative $\partial_x^\alpha f(g(x))$, where $f, g\colon \mathbb{R}^n \to \mathbb{R}^n$, $\alpha \in \mathbb{N}_0^n$ and $\partial_x^\alpha = \partial_{x_1}^{\alpha_1} \dots \partial_{x_n}^{\alpha_n}$. The formula will be shown via induction using the chain rule, it focusses on the structure which will be important in the sequel. We denote the respective functions by $f(y)$ and $g(x)$.

For $|\alpha| := \sum_{k=1}^n \alpha_k \ge 1$

$$\begin{aligned} \partial_x^\alpha f(g(x)) &= Df(g(x)) \partial_x^\alpha g(x) \\ &\quad + \sum_{2 \le |\beta| \le |\alpha|} (\partial_y^\beta f)(g(x))\, c_\beta^\alpha \prod_{j=1}^{|\beta|} \partial_x^{\gamma(\alpha,\beta,j)} g_{l(\alpha,\beta,j)}(x) \qquad (2.22) \\ &= \sum_{1 \le |\beta| \le |\alpha|} (\partial_y^\beta f)(g(x))\, c_\beta^\alpha \prod_{j=1}^{|\beta|} \partial_x^{\gamma(\alpha,\beta,j)} g_{l(\alpha,\beta,j)}(x) \qquad (2.23) \end{aligned}$$

holds with constants $c_\beta^\alpha \in \mathbb{R}$, where g_l denotes the l-th component of g and the multiindex $\gamma(\alpha, \beta, j)$ satisfies $1 \le |\gamma(\alpha, \beta, j)| \le |\alpha| - |\beta| + 1$. Note that (2.23) is a consequence of (2.22). Thus we show (2.22) via induction with respect to $|\alpha| \geq 1$.

For $|\alpha| = 1$ the sum is empty and (2.22) follows easily by the chain rule. Now assume that (2.22) holds for $|\alpha| \geq 1$. Let $|\alpha'| = |\alpha| + 1$ and write $\alpha' = \alpha + e_i$, where $e_i = (0, \ldots, 0, 1, 0, \ldots, 0)$ with the 1 at the i-th position. Then by induction we obtain

$$\begin{aligned}
&\partial_x^{e_i} \partial_x^\alpha f(g(x)) \\
&= \partial_x^{e_i} \left(\sum_{l=1}^{n} (\partial_y^{e_l} f)(g(x)) \partial_x^\alpha g_l(x) \right. \\
&\quad \left. + \sum_{2 \leq |\beta| \leq |\alpha|} (\partial_y^\beta f)(g(x))\, c_\beta^\alpha \prod_{j=1}^{|\beta|} \partial_x^\gamma g_{l(\alpha,\beta,j)}(x) \right) \\
&= Df(g(x)) \partial_x^{e_i} \partial_x^\alpha g(x) + \sum_{l=1}^{n} \sum_{k=1}^{n} (\partial_y^{e_l + e_k} f)(g(x)) \partial_x^\alpha g_l(x) \partial_x^{e_i} g_k(x) \\
&\quad + \sum_{2 \leq |\beta| \leq |\alpha|} \sum_{k=1}^{n} (\partial_y^{\beta + e_k} f)(g(x)) \partial_x^{e_i} g_k(x)\, c_\beta^\alpha \prod_{j=1}^{|\beta|} \partial_x^\gamma g_{l(\alpha,\beta,j)}(x) \\
&\quad + \sum_{2 \leq |\beta| \leq |\alpha|} (\partial_y^\beta f)(g(x))\, c_\beta^\alpha \prod_{j=1}^{|\beta|} \partial_x^{\gamma + e_i} g_{l(\alpha,\beta,j)}(x) \\
&= Df(g(x)) \partial_x^{\alpha'} g(x) + \sum_{2 \leq |\beta| \leq |\alpha|+1} (\partial_y^\beta f)(g(x))\, c_\beta^{\alpha'} \prod_{j=1}^{|\beta|} \partial_x^\gamma g_{l(\alpha',\beta,j)}(x),
\end{aligned}$$

where now $1 \leq |\gamma| \leq |\alpha| - |\beta| + 2$. This shows (2.22).

We have set $\epsilon = \frac{\nu}{3}$ and with this ϵ we define the matrix S as in Lemma 2.27. Set $\nu' := \frac{\nu}{3}$, i.e. $\nu - 2\epsilon = \nu'$ and $\tilde{\nu} = \frac{4}{3}\nu = 4\nu'$. Since $f \in C^1(\mathbb{R}^n, \mathbb{R}^n)$ we can choose r in $\tilde{K} := \tilde{K}_r^{\mathfrak{q}}(x_0)$ so small that

$$\|S[Df(x) - Df(x_0)]S^{-1}\| \leq \epsilon \tag{2.24}$$

holds for all $x \in \tilde{K}$. Now we show that

$$\|S\partial_x^\alpha (S_t x - x_0)\| \leq C_\alpha e^{-\nu' t} \tag{2.25}$$

holds for all α with $|\alpha| \leq \sigma$, $x \in \tilde{K}$ and $t \geq 0$. We prove this by induction with respect to $k = |\alpha|$.

For $k = 0$ the inequality follows from (2.21):

$$\|S(S_t x - x_0)\| \leq \|S\|\, c_2^{\frac{1}{2}} r e^{-\tilde{\nu}\frac{1}{2}t}.$$

Now assume that (2.25) holds for all $|\alpha| = k - 1 \geq 0$. Let $|\alpha'| = k$. Write $\alpha' = \alpha + e_i$, where $|\alpha| = k - 1$ and $e_i = (0, \ldots, 0, 1, 0, \ldots, 0)$ with the 1 at i-th position. We use (2.22) with $g(x) = S_t x$. Since $f \in C^\sigma$ and $|\alpha'| = k \leq \sigma$, we have for all $x \in \tilde{K}$ and $t \geq 0$

$$\begin{aligned}
&\|S\partial_x^{\alpha'} S_t x\| \cdot \frac{d}{dt}\|S\partial_x^{\alpha'} S_t x\| \\
&= \frac{1}{2}\frac{d}{dt}\|S\partial_x^{\alpha'} S_t x\|^2 \\
&= (S\partial_x^{\alpha'} S_t x)^T S(\partial_x^{\alpha'} \underbrace{\partial_t S_t x}_{= f(S_t x)}) \\
&= (S\partial_x^{\alpha'} S_t x)^T S\Bigg(Df(S_t x)\partial_x^{\alpha'} S_t x \\
&\quad + \sum_{2\leq|\beta|\leq k} (\partial_y^\beta f)(S_t x)\, c_\beta^{\alpha'} \prod_{j=1}^{|\beta|} \partial_x^{\gamma(\alpha',\beta,j)} (S_t x)_{l(\alpha',\beta,j)}\Bigg) \\
&\leq (S\partial_x^{\alpha'} S_t x)^T SDf(S_t x)S^{-1}(S\partial_x^{\alpha'} S_t x) + C\|S\partial_x^{\alpha'} S_t x\|e^{-2\nu' t}
\end{aligned}$$

since each summand of the sum is a product of at least two terms of the form $\partial_x^\gamma (S_t x)_{l(\alpha',\beta,j)}$ with $|\gamma| \leq |\alpha'| - 1 = |\alpha|$, and we can thus apply the induction assumption (2.25). Note that we could exchange $\partial_x^{\alpha'}$ and ∂_t since the solution $S_t x$ is smooth enough, cf. [36], Chapter V, Theorem 4.1. Hence,

$$\begin{aligned}
&\|S\partial_x^{\alpha'} S_t x\| \cdot \frac{d}{dt}\|S\partial_x^{\alpha'} S_t x\| \\
&\leq (S\partial_x^{\alpha'} S_t x)^T SDf(x_0)S^{-1}(S\partial_x^{\alpha'} S_t x) \\
&\quad + \|S[Df(S_t x) - Df(x_0)]S^{-1}\| \cdot \|S\partial_x^{\alpha'} S_t x\|^2 + C\|S\partial_x^{\alpha'} S_t x\|e^{-2\nu' t} \\
&\leq (-\nu + 2\epsilon)\|S\partial_x^{\alpha'} S_t x\|^2 + C\|S\partial_x^{\alpha'} S_t x\|e^{-2\nu' t} \\
&= -\nu'\|S\partial_x^{\alpha'} S_t x\|^2 + C\|S\partial_x^{\alpha'} S_t x\|e^{-2\nu' t}
\end{aligned}$$

by (2.24) and Lemma 2.27. Now we use Gronwall's Lemma for $\|S\partial_x^{\alpha'} S_t x\| = G(t) \geq 0$: if $\frac{d}{dt}G(t) \leq -\nu' G(t) + Ce^{-2\nu' t}$, then, by integration from 0 to t

$$\begin{aligned}
\frac{d}{d\tau}\left(e^{\nu'\tau}G(\tau)\right) &\leq Ce^{-\nu'\tau} \\
e^{\nu' t}G(t) - G(0) &\leq -\frac{C}{\nu'}\left(e^{-\nu' t} - 1\right) \\
G(t) &\leq e^{-\nu' t}\left[G(0) - \frac{C}{\nu'}\left(e^{-\nu' t} - 1\right)\right] \\
&\leq e^{-\nu' t}\left[G(0) + \frac{C}{\nu'}\right].
\end{aligned}$$

Thus, with $G(t) = \|S\partial_x^{\alpha'} S_t x\|$,

$$\|S\partial_x^{\alpha'} S_t x\| \leq C' e^{-\nu' t}$$

$$\text{and thus } \|\partial_x^{\alpha'} S_t x\| \leq C'' e^{-\nu' t}$$

with a uniform constant C'' for all $x \in \tilde{K}$. This shows (2.25).

Next, we show that $\int_0^\infty \partial_x^\alpha p(S_t x)\,dt$ converges uniformly with respect to x for $1 \leq |\alpha| \leq \sigma$. As above, let $x \in A(x_0)$ and let O be a bounded, open neighborhood of x, such that $\overline{O} \subset A(x_0)$. Since $\overline{O}$ is compact, Proposition 2.11 implies that there is a $T \in \mathbb{R}_0^+$ such that $S_{T+t}\overline{O} \subset \tilde{K}$ holds for all $t \geq 0$. Hence, it is sufficient to show that $\int_0^\infty \partial_x^\alpha p(S_t x)\,dt$ is uniformly bounded for all $x \in \tilde{K}$ by a similar argumentation as above. By a similar statement as (2.23), where we use $p\colon \mathbb{R}^n \to \mathbb{R}$ instead of $f\colon \mathbb{R}^n \to \mathbb{R}^n$, and $g(x) = S_t x$, we have

$$\partial_x^\alpha p(S_t x) = \sum_{1\leq|\beta|\leq|\alpha|} (\partial_y^\beta p)(S_t x)\, c_\beta^\alpha \prod_{j=1}^{|\beta|} \partial_x^\gamma (S_t x)_{l(\alpha,\beta,j)}$$

where the multiindex γ satisfies $1 \leq |\gamma| \leq |\alpha| - |\beta| + 1$. Hence, by (2.25),

$$\int_0^\infty \partial_x^\alpha p(S_t x)\,dt \leq \sum_{1\leq|\beta|\leq|\alpha|} \max_{x\in\tilde{K}} |\partial_x^\beta p(x)| \tilde{C} \int_0^\infty e^{-\nu' t}\,dt$$

which is uniformly bounded for all $x \in \tilde{K}$. This proves $V \in C^\sigma$.

Now assume that $f(x)$ is bounded, i.e. there is a constant $C > 0$ such that $\|f(x)\| \leq C$ holds for all $x \in A(x_0)$. If for an $R \geq 0$ the set

$$K_R := \{x \in A(x_0) \mid V(x) \leq R^2\}$$

is not compact in $\mathbb{R}^n$, then either K_R is unbounded or not closed.

We will first show that all sets K_R are bounded: Indeed, let $\mathfrak{q}$ be a local Lyapunov function with local Lyapunov basin $\tilde{K} = \tilde{K}_r^{\mathfrak{q}}(x_0)$. Denote the non-characteristic hypersurface $\partial\tilde{K} = \Omega$ and let $\theta \in C^\sigma(A(x_0) \setminus \{x_0\}, \mathbb{R})$ be the function satisfying $S_{\theta(x)} x \in \Omega$, cf. Theorem 2.38. Then for $x \in A(x_0) \setminus \tilde{K}$ we have $\theta(x) > 0$ and

$$\begin{aligned} x - S_{\theta(x)} x &= -\int_0^{\theta(x)} f(S_\tau x)\,d\tau \\ \|x\| &\leq \|S_{\theta(x)} x\| + |\theta(x)| \sup_{\xi\in A(x_0)} \|f(\xi)\| \\ &\leq S + \theta(x) \cdot C, \end{aligned} \tag{2.26}$$

where $S := \max_{\xi\in\Omega} \|\xi\|$. By assumption, there is a constant $c > 0$ such that $p(x) \geq c$ holds for all $x \notin \tilde{K}$. Let $x \in K_R \setminus \tilde{K}$, i.e. $\theta(x) > 0$. Then

$$\begin{aligned} R^2 &\geq V(x) \\ &= \int_0^{\theta(x)} p(S_t x)\,dt + \int_{\theta(x)}^\infty p(S_t x)\,dt \\ &\geq \theta(x) c \\ &\geq \frac{c\,(\|x\| - S)}{C} \end{aligned}$$

by (2.26). Hence, $K_R \subset \tilde{K} \cup \overline{B_{\frac{CR^2}{c}+S}(0)}$ and, in particular, K_R is bounded.

Now assume that K_R is not closed for an R: Then there is a sequence $x_k \in K_R$ with $\lim_{k\to\infty} x_k = x \notin K_R$. By definition of K_R, $x \notin A(x_0)$.

Let $\epsilon > 0$ be such that $B_\epsilon(x_0) \subset A(x_0)$. Choose the constant $c > 0$ such that $p(x) \geq c$ for all $x \notin B_{\frac{\epsilon}{2}}(x_0)$. Set $T_0 := \frac{R^2+1}{c} > 0$. We have $S_t x \notin B_\epsilon(x_0)$ for all $t \in [0, T_0]$, since $B_\epsilon(x_0) \subset A(x_0)$. By continuity of $S_t x$ with respect to t and x there is a $\delta > 0$ such that $\|S_t x - S_t y\| < \frac{\epsilon}{2}$ holds for all y with $\|x - y\| < \delta$ and for all $t \in [0, T_0]$. There is a $k \in \mathbb{N}$ such that $y = x_k$ satisfies $\|x - x_k\| < \delta$. For this point we thus have $S_t x_k \notin B_{\frac{\epsilon}{2}}(x_0)$ for all $t \in [0, T_0]$. Hence,

$$V(x_k) \geq \int_0^{T_0} p(S_t x_k)\, dt \geq T_0 \cdot c = R^2 + 1,$$

which is a contradiction to $x_k \in K_R$. □

Remark 2.47 *The definition of V and the convergence of the integral was shown by Zubov [70]. He assumes that f is continuous and shows that also V is continuous. Aulbach [3] showed that V is analytic, if f is analytic.*

Proposition 2.48 *A function $V \in C^\sigma(A(x_0), \mathbb{R})$ with $V'(x) = -p(x)$, where p is as in Theorem 2.46, is unique up to a constant.*

PROOF: Assume, $V_1, V_2 \in C^\sigma(A(x_0), \mathbb{R})$ both satisfy $V_1'(x) = V_2'(x) = -p(x)$. Let $V_1(x_0) = c_1$ and $V_2(x_0) = c_2$ and define $C := c_1 - c_2$.

Fix $x \in A(x_0) \setminus \{x_0\}$. For $t \geq 0$ we have $V_i(S_t x) = V_i(x) + \int_0^t V_i'(S_\tau x)\, d\tau = V_i(x) - \int_0^t p(S_\tau x)\, d\tau$, where $i = 1, 2$. Thus, $V_1(S_t x) - V_2(S_t x) = V_1(x) - V_2(x)$. Since $\lim_{t\to\infty} S_t x = x_0$ and V_1 and V_2 are continuous, we have $V_1(x) - V_2(x) = \lim_{t\to\infty}[V_1(S_t x) - V_2(S_t x)] = V_1(x_0) - V_2(x_0) = c_1 - c_2 = C$ and thus $V_1(x) = V_2(x) + C$ for all $x \in A(x_0)$. □

Remark 2.49 *The function V of Theorem 2.46 is unique if we claim $V(x_0) = 0$, cf. Proposition 2.48. Unless stated otherwise, we will denote by V the unique function satisfying $V(x_0) = 0$ and $V'(x) = -p(x)$ for all $x \in A(x_0)$.*

Corollary 2.50 *For linear systems and $p(x) = \|x - x_0\|^2$, we have $V(x) = \mathfrak{v}(x)$, where V is defined in Remark 2.49 and $\mathfrak{v}$ is as in Remark 2.34. Indeed, since the functions satisfy $V'(x) = \mathfrak{v}'(x) = -\|x - x_0\|^2$ and $V(x_0) = \mathfrak{v}(x_0) = 0$, they are equal by Proposition 2.48.*

If we also want to fix the values of V on a non-characteristic hypersurface, then the resulting function V^* is not defined in x_0 in general, similarly to the function T.

Proposition 2.51 (Existence of V^*) *Let $\dot{x} = f(x)$, $f \in C^\sigma(\mathbb{R}^n, \mathbb{R}^n)$, $\sigma \geq 1$. Let x_0 be an equilibrium such that $-\nu < 0$ is the maximal real part of all eigenvalues of $Df(x_0)$. Let Ω be a non-characteristic hypersurface and $p(x)$ a function as in Theorem 2.46.*

For all functions $H \in C^\sigma(\Omega, \mathbb{R})$ there is a function $V^* \in C^\sigma(A(x_0) \setminus \{x_0\}, \mathbb{R})$ satisfying

$$\begin{aligned}(V^*)'(x) &= -p(x) \text{ for all } x \in A(x_0) \setminus \{x_0\}, \\ V^*(x) &= H(x) \text{ for all } x \in \Omega.\end{aligned}$$

If $\sup_{x \in A(x_0)} \|f(x)\| < \infty$, then

$$K_R^* := \{x \in A(x_0) \setminus \{x_0\} \mid V^*(x) \leq R^2\} \cup \{x_0\}$$

is a compact set in $\mathbb{R}^n$ for all $R \geq 0$.

PROOF: Define

$$V^*(x) = V(x) + H(S_{\theta(x)}x) - V(S_{\theta(x)}x)$$

for $x \in A(x_0) \setminus \{x_0\}$ where V is the function of Theorem 2.46 with $V(x_0) = 0$ and θ is defined in Theorem 2.38. Then for the orbital derivative $(V^*)'(x) = V'(x) = -p(x)$ holds. For $x \in \Omega$, $V^*(x) = V(x) + H(x) - V(x) = H(x)$. Since $K_R^* \subset \{x \in A(x_0) \mid V(x) \leq \max(R^2 + \max_{\xi \in \Omega}[V(\xi) - H(\xi)], 0)\}$, K_R^* is a closed subset of a compact set by Theorem 2.46 and thus compact for each $R \geq 0$. □

2.3.4 Taylor Polynomial of *V*

V is a smooth function in $A(x_0)$. In particular, V is C^σ at x_0 and we can study its Taylor polynomial at x_0. The Taylor polynomial will also be of use for the approximation. The reason is that approximating the function V by an approximation v using radial basis functions, we obtain an error estimate $|V'(x) - v'(x)| \leq \iota$ which implies $v'(x) \leq V'(x) + \iota = -p(x) + \iota$. Thus, $v'(x) \leq -p(x) + \iota < 0$ only where $p(x) > \iota$. Thus, we cannot guarantee that v has negative orbital derivative in the neighborhood of x_0, where $p(x) \leq \iota$. This may just be a lack of estimates, but we will also show that typically there are indeed points near x_0 where $v'(x) > 0$. Hence, this local problem has to be solved, cf. Section 4.2. One possibility, which will be discussed in Section 4.2.3, uses the Taylor polynomial of V. In this section we will derive a method to calculate the Taylor polynomial of V without explicitly knowing V.

We consider the function V, which satisfies $V'(x) = -\|x - x_0\|^2$, i.e. $p(x) = \|x - x_0\|^2$. We will describe a way to calculate the Taylor polynomial $\mathfrak{h}$ of V explicitly. Moreover, we will construct a function $\mathfrak{n}$ with $\mathfrak{n}(x) = \mathfrak{h}(x) + M\|x - x_0\|^{2H}$ which satisfies $\mathfrak{n}(x) > 0$ for all $x \neq x_0$. With this function we define

$$W(x) := \frac{V(x)}{\mathfrak{n}(x)}$$

for $x \neq x_0$ and $W(x_0) = 1$, and show several properties of W in Proposition 2.58. In particular, W turns out to be C^{P-2} in x_0, where P is the order of the Taylor polynomial.

We are also interested in a method to calculate the Taylor polynomial. For linear systems, we have $V(x) = \mathfrak{v}(x)$, cf. Corollary 2.50. For nonlinear systems, $\mathfrak{v}$ is the Taylor polynomial of V of order two. Note that $\mathfrak{v}$ is characterized by

$$\begin{aligned}&\langle \nabla \mathfrak{v}(x), Df(x_0)(x - x_0)\rangle = -\|x - x_0\|^2\\ \Longleftrightarrow\ &\mathfrak{v}'(x) = \langle \nabla \mathfrak{v}(x), f(x)\rangle = -\|x - x_0\|^2 + o(\|x - x_0\|^2)\end{aligned}$$

Note that $\nabla \mathfrak{v}(x) = O(\|x - x_0\|)$, cf. Remark 2.34, and $f(x) - Df(x_0)(x - x_0) = o(\|x - x_0\|)$.

The second equation can be used to explicitly calculate $\mathfrak{v}$. We generalize this idea: The local Lyapunov function $\mathfrak{v}$ satisfies $\mathfrak{v}'(x) = -\|x - x_0\|^2 + o(\|x - x_0\|^2)$. We define a function $\mathfrak{h}$ which is a polynomial in $(x - x_0)$ of order $P \geq 2$ and satisfies $\mathfrak{h}'(x) = -\|x - x_0\|^2 + o(\|x - x_0\|^P)$; note that for $P = 2$, $\mathfrak{h} = \mathfrak{v}$. The function $\mathfrak{h}$ turns out to be the Taylor polynomial of V of order P. Later, in order to obtain a function $\mathfrak{n}(x) > 0$ for all $x \neq x_0$, we add another polynomial of high order, i.e. $\mathfrak{n}(x) = \mathfrak{h}(x) + M\|x - x_0\|^{2H}$, cf. Definition 2.56.

We start by defining the function $\mathfrak{h}$. One can calculate the constants c_α in (2.27) by solving (2.28) where f is replaced by its Taylor polynomial of degree $P - 1$. An example will be given in Example 2.55.

Definition 2.52 (Definition of $\mathfrak{h}$). *Let $\dot{x} = f(x)$, $f \in C^\sigma(\mathbb{R}^n, \mathbb{R}^n)$, $\sigma \geq 1$. Let x_0 be an equilibrium such that the real parts of all eigenvalues of $Df(x_0)$ are negative. Let $\sigma \geq P \geq 2$. Let $\mathfrak{h}$ be the function*

$$\mathfrak{h}(x) := \sum_{2 \leq |\alpha| \leq P} c_\alpha (x - x_0)^\alpha, \tag{2.27}$$

$c_\alpha \in \mathbb{R}$, such that

$$\mathfrak{h}'(x) = \langle \nabla \mathfrak{h}(x), f(x)\rangle = -\|x - x_0\|^2 + \varphi(x), \tag{2.28}$$

where $\varphi(x) = o(\|x - x_0\|^P)$.

Lemma 2.53. *There is one and only one function $\mathfrak{h}$ of the form (2.27) which satisfies (2.28).*

The function $\mathfrak{h}$ is the Taylor polynomial of V (with $V'(x) = -\|x - x_0\|^2$ and $V(x_0) = 0$) of order P.

The proof of the lemma will be given below. First, we explain in Remark 2.54 and Example 2.55 how to calculate $\mathfrak{h}$ explicitly. This reflects at the same time the main idea for the proof of Lemma 2.53.

Remark 2.54 *The equation (2.28) can be solved by plugging the ansatz (2.27) in (2.28) and replacing f by its Taylor polynomial of order $P - 1$. Then (2.28) becomes*

$$\left\langle \sum_{2 \leq |\alpha| \leq P} c_\alpha \nabla (x - x_0)^\alpha, Df(x_0)(x - x_0) + \sum_{2 \leq |\beta| \leq P-1} \frac{\partial^\beta f(x_0)}{\beta!}(x - x_0)^\beta \right\rangle$$
$$= -\|x - x_0\|^2 + o(\|x - x_0\|^P). \tag{2.29}$$

The difference of f to its Taylor polynomial is of order $o(\|x-x_0\|^{P-1})$ and by multiplication with $\nabla\mathfrak{h}(x) = O(\|x-x_0\|)$ the remaining term is of order $o(\|x-x_0\|^P)$.

Example 2.55 *Consider the differential equation*

$$\begin{cases} \dot{x} = -x + x^3 \\ \dot{y} = -\frac{1}{2}y + x^2 \end{cases} \tag{2.30}$$

with equilibrium $x_0 = (0,0)$. We calculate the polynomial $\mathfrak{h}(x,y)$ and start with the terms $\mathfrak{h}_2(x,y)$ of second order. We have shown that $\mathfrak{h}_2(x,y) = \mathfrak{v}(x,y) = (x,y)B\begin{pmatrix} x \\ y \end{pmatrix}$. B is the solution of $Df(0,0)^T B + BDf(0,0) = -I$, cf. Remark 2.34, in our case $B = \begin{pmatrix} \frac{1}{2} & 0 \\ 0 & 1 \end{pmatrix}$. Thus $\mathfrak{h}_2(x,y) = \frac{1}{2}x^2 + y^2$.

For the terms of order three, which we denote by $\mathfrak{h}_3(x,y)$, we calculate the left-hand side of (2.29) with $P = 3$:

$$\begin{aligned} &\left\langle \nabla\left[\mathfrak{h}_2(x,y) + \mathfrak{h}_3(x,y)\right], \begin{pmatrix} -x \\ -\frac{1}{2}y \end{pmatrix} + \begin{pmatrix} 0 \\ x^2 \end{pmatrix} \right\rangle \\ &= -x^2 - y^2 + 2x^2y + \left\langle \nabla\mathfrak{h}_3(x,y), \begin{pmatrix} -x \\ -\frac{1}{2}y \end{pmatrix} \right\rangle + o(\|(x,y)\|^3). \end{aligned}$$

Thus, $\mathfrak{h}_3(x,y) = ax^2y$, where $2 - 2a - \frac{1}{2}a = 0$, i.e. $\mathfrak{h}_3(x,y) = \frac{4}{5}x^2y$.

For the terms of order four, which we denote by $\mathfrak{h}_4(x,y)$, we calculate the left-hand side of (2.29) with $P = 4$:

$$\begin{aligned} &\left\langle \nabla\left[\mathfrak{h}_2(x,y) + \mathfrak{h}_3(x,y) + \mathfrak{h}_4(x,y)\right], \begin{pmatrix} -x \\ -\frac{1}{2}y \end{pmatrix} + \begin{pmatrix} 0 \\ x^2 \end{pmatrix} + \begin{pmatrix} x^3 \\ 0 \end{pmatrix} \right\rangle \\ &= -x^2 - y^2 + x^4 + \frac{4}{5}x^4 + \left\langle \nabla\mathfrak{h}_4(x,y), \begin{pmatrix} -x \\ -\frac{1}{2}y \end{pmatrix} \right\rangle + o(\|(x,y)\|^4). \end{aligned}$$

Thus, $\mathfrak{h}_4(x,y) = bx^4$, where $1 + \frac{4}{5} - 4b = 0$, i.e. $\mathfrak{h}_4(x,y) = \frac{9}{20}x^4$.

For the terms of order five, which we denote by $\mathfrak{h}_5(x,y)$, we calculate the left-hand side of (2.29) with $P = 5$:

$$\begin{aligned} &\left\langle \nabla\sum_{i=2}^{5}\mathfrak{h}_i(x,y), \begin{pmatrix} -x \\ -\frac{1}{2}y \end{pmatrix} + \begin{pmatrix} 0 \\ x^2 \end{pmatrix} + \begin{pmatrix} x^3 \\ 0 \end{pmatrix} \right\rangle \\ &= -x^2 - y^2 + \frac{8}{5}x^4y + \left\langle \nabla\mathfrak{h}_5(x,y), \begin{pmatrix} -x \\ -\frac{1}{2}y \end{pmatrix} \right\rangle + o(\|(x,y)\|^5). \end{aligned}$$

Thus, $\mathfrak{h}_4(x,y) = cx^4y$, where $\frac{8}{5} - 4c - \frac{1}{2}c = 0$, i.e. $\mathfrak{h}_5(x,y) = \frac{16}{45}x^4y$.

Altogether, the function $\mathfrak{h}$ for $P = 5$ is thus given by

$$\mathfrak{h}(x,y) = \underbrace{\frac{1}{2}x^2 + y^2}_{=\mathfrak{v}(x,y)} + \frac{4}{5}x^2y + \frac{9}{20}x^4 + \frac{16}{45}x^4y. \tag{2.31}$$

A corresponding function $\mathfrak{n}$ *will be given in Example 2.57. The example will further be considered in Example 6.3.*

PROOF: [of Lemma 2.53] By Remark 2.54, (2.28) turns out to be a system of linear equations for c_α when considering each order of $(x - x_0)$. We will show that this system has a unique solution.

We will first show, that it is sufficient to consider the special case when $Df(x_0) = J$ is an upper diagonal matrix. Indeed, there is an invertible matrix S such that $J = SDf(x_0)S^{-1}$ is the (complex) Jordan normal form of $Df(x_0)$. In particular, J is an upper diagonal matrix. Note that S and J are complex-valued matrices. However, if we can show that there is a unique solution $\mathfrak{h}$, it is obvious that all coefficients c_α are in fact real. Define $y := S(x-x_0)$, then $x = S^{-1}y+x_0$. The equation $\dot{x} = f(x)$ then is equivalent to $\dot{y} = Sf(S^{-1}y+x_0) =: g(y)$, and we have $Dg(y) = SDf(S^{-1}y+x_0)S^{-1}$ and $Dg(0) = SDf(x_0)S^{-1} = J$. We will later show that the special case $\dot{y} = g(y)$ with upper diagonal matrix $Dg(0) = J$ has a unique solution $\mathfrak{h}(y) = \sum_{2\le|\alpha|\le P} c_\alpha y^\alpha$ of the following equation, which is similar to (2.28):

$$[\nabla_y \mathfrak{h}(y)]^T g(y) = -y^T C y + o(\|y\|^P), \tag{2.32}$$

where $C = (S^{-1})^T S^{-1}$.

Note that $\mathfrak{h}(y) = \sum_{2\le|\alpha|\le P} c_\alpha y^\alpha$ is a solution of (2.32) if and only if $\tilde{\mathfrak{h}}(x) := \sum_{2\le|\alpha|\le P} c_\alpha (S(x-x_0))^\alpha = \sum_{2\le|\alpha|\le P} c'_\alpha (x-x_0)^\alpha$ is a solution of (2.28), since we have with $y = S(x-x_0)$

$$\begin{aligned} \tilde{\mathfrak{h}}'(x) &= \left[\nabla_x \tilde{\mathfrak{h}}(x)\right]^T f(x) \\ &= [\nabla_y \mathfrak{h}(y)]^T SS^{-1} g(y) \\ \text{and } -y^T C y + o(\|y\|^P) &= -\|S^{-1}y\|^2 + o(\|S^{-1}y\|^P) \\ &= -\|x-x_0\|^2 + o(\|x-x_0\|^P). \end{aligned}$$

Hence, we have to show that there exists a unique solution $\mathfrak{h}(y) = \sum_{2\le|\alpha|\le P} c_\alpha y^\alpha$ of (2.32), i.e.

$$\left\langle \sum_{2\le|\alpha|\le P} c_\alpha \nabla y^\alpha, Jy + \sum_{2\le|\beta|\le P-1} \frac{\partial^\beta g(0)}{\beta!} y^\beta \right\rangle = -y^T C y + o(\|y\|^P) \tag{2.33}$$

where J is an upper diagonal matrix such that all eigenvalues have negative real parts, C is a symmetric matrix and $P \ge 2$. We consider the terms order by order in y. The lowest appearing order is two and the terms of this order in y of both sides of (2.33) are

$$\left\langle \nabla \sum_{|\alpha|=2} c_\alpha y^\alpha, Jy \right\rangle = -y^T C y. \tag{2.34}$$

Writing the terms of order two $\sum_{|\alpha|=2} c_\alpha y^\alpha = y^T By$, (2.34) becomes $J^T B + BJ = -C$. Since the eigenvalues of J have negative real parts, this equation has a unique solution B; the proof for the complex-valued matrix J is the same as for the real case, cf. Theorem 2.30 and [57], Lemma 4.6.14, p. 167.

Now we show by induction with respect to $|\alpha| \leq P$ that the constants c_α are uniquely determined by (2.33): For $|\alpha| = 2$ this has just been done. Now let $P \geq |\alpha| = k \geq 3$. We have $\mathfrak{h}(y) = \sum_{2\leq|\alpha|\leq k-1} c_\alpha y^\alpha + \sum_{|\alpha|=k} c_\alpha y^\alpha$. The constants c_α with $2 \leq |\alpha| \leq k-1$ are fixed. We will show that there is a unique solution for the constants c_α with $|\alpha| = k$ such that $\langle \nabla \mathfrak{h}(y), Jy + \sum_{2\leq|\beta|\leq P-1} \frac{\partial^\beta g(0)}{\beta!} y^\beta \rangle + y^T Cy = o(\|y\|^k)$, i.e. the expression has no terms of order $\leq k$. Consider (2.33): the terms of order $\leq k-1$ satisfy the equation by induction. Now consider the terms of order $|\alpha| = k$:

$$\left\langle \sum_{|\alpha|=k} c_\alpha \nabla y^\alpha, Jy \right\rangle = -\left\langle \sum_{2\leq|\alpha|\leq k-1, |\beta|=k-|\alpha|+1} c_\alpha \nabla y^\alpha, \frac{\partial^\beta g(0)}{\beta!} y^\beta \right\rangle.$$

Since all c_α on the right-hand side are known and all c_α on the left-hand side are unknown, this is equivalent to an inhomogeneous system of linear equations. It has a unique solution, if and only if the corresponding homogeneous system has only the zero solution. Therefore, we study the problem

$$\left\langle \sum_{|\alpha|=k} c_\alpha \nabla y^\alpha, Jy \right\rangle = 0, \tag{2.35}$$

which is equivalent to the corresponding homogeneous problem, and show that $c_\alpha = 0$ is its only solution. Note that $\operatorname{Re} J_{ii} < 0$ holds for all $1 \leq i \leq n$ since J is an upper diagonal matrix and J is the Jordan normal form of $Df(x_0)$, the eigenvalues of which have negative real parts. All terms of (2.35) are polynomials of order k.

We prove by induction that all coefficients c_α vanish. We introduce an order on $\tilde{A} := \{\alpha \in \mathbb{N}_0^n \mid \alpha_i \in \{0, \ldots, k\} \text{ for all } i \in \{1, \ldots, n\}\}$. Note that $\{\alpha \in \mathbb{N}_0^n \mid |\alpha| = k\} \subset \tilde{A}$. The order $\|\alpha\|$ on $\tilde{A}$ is such that α is the $(k+1)$-adic expansion of $\|\alpha\|$, i.e.

$$\mathbb{N}_0 \ni \|\alpha\| = \sum_{l=1}^{n} \alpha_l (k+1)^{l-1}.$$

Now we start the induction with respect to $\|\alpha\|$. The minimal α with $|\alpha| = k$ is $\alpha = (k, 0, \ldots, 0)$. The coefficient of y_1^k in (2.35) is $c_{(k,0,\ldots,0)} k J_{11}$ and since $J_{11} \neq 0$, we have $c_{(k,0,\ldots,0)} = 0$.

Now we assume that all coefficients c_α with $|\alpha| = k$ and $\|\alpha\| \leq A^*$ for some $A^* \in \mathbb{N}$ are zero. Let β be minimal with $\|\beta\| > A^*$ and $|\beta| = k$. We will show that $c_\beta = 0$. Consider the coefficient of y^β in (2.35). How can we obtain such terms in (2.35)? We consider the terms of $\sum_{|\alpha|=k} c_\alpha y^\alpha$ with $\alpha_i = \beta_i + 1$, $\alpha_j = \beta_j - 1$ and $\alpha_l = \beta_l$ for $l \in \{1, \ldots, n\} \setminus \{i, j\}$. Note that $i = j$ is possible.

These terms in $\sum_{|\alpha|=k} c_\alpha y^\alpha$ result in a term y^β in (2.35) with coefficient $(\beta_i+1)c_\alpha J_{ij}$. Altogether, the coefficient of y^β in (2.35) is

$$\sum_{j=1,\beta_j\neq 0}^{n}\left(\sum_{i=1,i\neq j,\beta_i\neq k}^{n} c_{\beta+e_{ij}}(\beta_i+1)J_{ij}+c_\beta\beta_j J_{jj}\right)=0, \tag{2.36}$$

where $e_{ij}=(0,\ldots,0,1,0,\ldots,0,-1,0,\ldots,0)$ with the 1 at i-th position and the -1 at j-th position.

Note that $J_{ij}=0$ for $j<i$ since J is an upper diagonal matrix. Moreover, $c_{\beta+e_{ij}}=0$ for $j>i$ by induction since $\|\beta+e_{ij}\|<\|\beta\|$. Indeed,

$$\|\beta+e_{ij}\|-\|\beta\|=(k+1)^{i-1}-(k+1)^{j-1}<0$$

since $j>i$. Hence, (2.36) becomes

$$c_\beta\sum_{j=1,\beta_j\neq 0}^{n}\beta_j J_{jj}=0.$$

Since the real part of all J_{jj} is negative, $\beta_j>0$ and the sum is not empty because $|\beta|=k$, we can conclude $c_\beta=0$.

We show that $\mathfrak{h}$ is the Taylor polynomial of V of order P: Let $V(x)=\sum_{0\leq|\alpha|\leq P} d_\alpha(x-x_0)^\alpha+\tilde{\varphi}(x)$ with $\tilde{\varphi}(x)=o(\|x-x_0\|^P)$, i.e. $\sum_{0\leq|\alpha|\leq P} d_\alpha(x-x_0)^\alpha$ is the Taylor polynomial of V of order P. Then

$$\begin{aligned}-\|x-x_0\|^2&=V'(x)\\&=\langle\nabla\sum_{0\leq|\alpha|\leq P} d_\alpha(x-x_0)^\alpha,f(x)\rangle+\underbrace{\langle\nabla\tilde{\varphi}(x),f(x)\rangle}_{=o(\|x-x_0\|^P)}\end{aligned} \tag{2.37}$$

since $\nabla\tilde{\varphi}(x)=o(\|x-x_0\|^{P-1})$, cf. Lemma 2.59, and $f(x)=O(\|x-x_0\|)$. By the uniqueness of $\mathfrak{h}$, $\sum_{0\leq|\alpha|\leq P} d_\alpha(x-x_0)^\alpha=\mathfrak{h}(x)$, since it satisfies (2.28). Note that $V(x_0)=0$ and the terms of order one must be zero by (2.37). □

The terms of $\mathfrak{h}$ of order two are the function $\mathfrak{v}$. $\mathfrak{v}$ is positive definite by Theorem 2.30. We add a higher order polynomial $M\|x-x_0\|^{2H}$ to $\mathfrak{h}$ in order to obtain a function $\mathfrak{n}$ for which $\mathfrak{n}(x)>0$ holds for all $x\neq x_0$.

Definition 2.56 (Definition of $\mathfrak{n}$). *Let $\sigma\geq P\geq 2$ and let $\mathfrak{h}$ be as in Definition 2.52. Let $H:=\lfloor\frac{P}{2}\rfloor+1$ and $M\geq 0$, and define*

$$\begin{aligned}\mathfrak{n}(x)&=\mathfrak{h}(x)+M\|x-x_0\|^{2H}\\&=\sum_{2\leq|\alpha|\leq P} c_\alpha(x-x_0)^\alpha+M\|x-x_0\|^{2H}.\end{aligned} \tag{2.38}$$

Choose the constant M so large, that $\mathfrak{n}(x)>0$ holds for all $x\neq x_0$.

PROOF: We show that the choice of the constant M is possible: Define the function $\tilde{\mathfrak{h}}(x) := \mathfrak{h}(x) + \|x - x_0\|^{2H}$. Assume that there is a point $\xi \neq x_0$ such that $\tilde{\mathfrak{h}}(\xi) \leq 0$; if there is no such point, then choose $M = 1$. Note that there are constants $0 < r \leq R$ such that $\tilde{\mathfrak{h}}(x) > 0$ holds for all $\|x - x_0\| \notin \{0\} \cup (r, R)$, since both the lowest and the highest order terms of $\tilde{\mathfrak{h}}$ are positive definite. Set $M := \underbrace{-\frac{\min_{\|x-x_0\|\in[r,R]} \tilde{\mathfrak{h}}(x)}{r^{2H}}}_{<0} + 2 > 1$. Then $\mathfrak{n}(x)$ as in (2.38) satisfies $\mathfrak{n}(x) > 0$ for all $\|x - x_0\| \notin \{0\} \cup (r, R)$ as above. For $\|x - x_0\| \in (r, R)$ we have

$$
\begin{aligned}
\mathfrak{n}(x) &= \tilde{\mathfrak{h}}(x) - \frac{\min_{\|x-x_0\|\in[r,R]} \tilde{\mathfrak{h}}(x)}{r^{2H}} \|x - x_0\|^{2H} + \|x - x_0\|^{2H} \\
&\geq \min_{\|x-x_0\|\in[r,R]} \tilde{\mathfrak{h}}(x)(1 - 1) + \|x - x_0\|^{2H} \\
&> 0.
\end{aligned}
$$

□

Example 2.57 *We consider again the differential equation (2.30) of Example 2.55. A function $\mathfrak{n}(x, y)$ for $P = 5$ is given by $\mathfrak{n}(x, y) = \mathfrak{h}(x, y) + (x^2 + y^2)^3$, i.e., cf. (2.31)*

$$
\begin{aligned}
\mathfrak{n}(x, y) &= \underbrace{\frac{1}{2}x^2 + y^2}_{=\mathfrak{v}(x,y)} + \frac{4}{5}x^2 y + \frac{9}{20}x^4 + \frac{16}{45}x^4 y + (x^2 + y^2)^3. \\
\mathfrak{n}'(x, y) &= -x^2 - y^2 \\
&\quad - \frac{173}{45}x^6 - 15x^4 y^2 - 12x^2 y^4 - 3y^6 + \frac{334}{45}x^6 y \\
&\quad + 12x^4 y^3 + 6y^5 x^2 + 6x^8 + 12x^6 y^2 + 6x^4 y^4.
\end{aligned}
$$

$\mathfrak{n}(x, y) > 0$ holds for $(x, y) \neq (0, 0)$ since $\frac{4}{5}x^2 y \geq -\frac{2}{5}(x^4 + y^2)$ and thus

$$
\begin{aligned}
\mathfrak{n}(x, y) &\geq \frac{1}{2}x^2 + \frac{3}{5}y^2 + \frac{1}{20}x^4 + \frac{16}{45}x^4 y + 3x^4 y^2 \\
&= \frac{1}{2}x^2 + \frac{3}{5}y^2 + 3x^4 \underbrace{\left(\frac{1}{60} + \frac{16}{135}y + y^2\right)}_{\geq 0}
\end{aligned}
$$

The example will further be considered in Example 6.3.

In Proposition 2.58 we show some properties of $\mathfrak{n}(x)$ and $W(x) := \frac{V(x)}{\mathfrak{n}(x)}$.

Proposition 2.58 *Let $\sigma \geq P \geq 2$ and let $\mathfrak{n}$ be as in Definition 2.56. Then*

1. *$\mathfrak{n}(x) > 0$ holds for all $x \neq x_0$.*

2. *For a compact set K, there is a $C > 0$ such that $\mathfrak{n}(x) \leq C\|x - x_0\|^2$ holds for all $x \in K$.*
3. *Set $W(x) = \frac{V(x)}{\mathfrak{n}(x)}$ for $x \in A(x_0) \setminus \{x_0\}$ and $W(x_0) = 1$. We have $W \in C^{P-2}(A(x_0), \mathbb{R})$ and $\partial_x^\alpha W(x_0) = 0$ for all $1 \leq |\alpha| \leq P - 2$.*

PROOF: 1. is clear by construction, cf. Definition 2.56.

For 2. note that $\mathfrak{n}(x) = \mathfrak{v}(x) + \sum_{3\leq|\alpha|\leq P} c_\alpha(x - x_0)^\alpha$ and $\mathfrak{v}(x) = (x - x_0)^T B(x - x_0)$ with a symmetric, positive definite matrix B. Denote the minimal and maximal eigenvalue of B by $0 < \lambda \leq \Lambda$, respectively, such that $\lambda\|x - x_0\|^2 \leq \mathfrak{v}(x) \leq \Lambda\|x - x_0\|^2$. There is an open neighborhood $U \subset B_1(x_0)$ of x_0 such that $\left|\sum_{3\leq|\alpha|\leq P} c_\alpha(x - x_0)^\alpha\right| \leq \frac{\lambda}{2}\|x - x_0\|^2$ holds for all $x \in U$. Hence,

$$\frac{\lambda}{2}\|x - x_0\|^2 \leq \mathfrak{n}(x) \leq \left(\Lambda + \frac{\lambda}{2}\right)\|x - x_0\|^2 \tag{2.39}$$

holds for all $x \in U$. If $K \subset U$, then set $C := \Lambda + \frac{\lambda}{2} > 0$. Otherwise define $M := \max_{x\in K\setminus U} \mathfrak{n}(x) > 0$ for the compact set $K\setminus U$ and $M_0 := \min_{x\in K\setminus U} \|x - x_0\|^2 > 0$. Set $C := \max\left(\Lambda + \frac{\lambda}{2}, \frac{M}{M_0}\right)$.

To prove 3., note that $V(x) \in C^\sigma(A(x_0), \mathbb{R})$ and $\mathfrak{n}(x) \in C^\infty(\mathbb{R}^n, \mathbb{R})$. Because of 1., the only point to consider is x_0. We first prove the following Lemmas 2.59 and 2.60.

Lemma 2.59. *Let $g \in C^s(\mathbb{R}^n, \mathbb{R})$.*

1. *Let $g(x) = o(\|x\|^p)$, where $p \leq s$, and $\alpha \in \mathbb{N}_0^n$ with $|\alpha| \leq p$. Then $\partial^\alpha g(x) \in C^{s-|\alpha|}(\mathbb{R}^n, \mathbb{R})$ and $\partial^\alpha g(x) = o(\|x\|^{p-|\alpha|})$.*
2. *Let $g(x) = O(\|x\|^p)$, where $p \leq s$, and $\alpha \in \mathbb{N}_0^n$ with $|\alpha| \leq p$. Then $\partial^\alpha g(x) \in C^{s-|\alpha|}(\mathbb{R}^n, \mathbb{R})$ and $\partial^\alpha g(x) = O(\|x\|^{p-|\alpha|})$.*

PROOF: We prove 1.: The first statement follows by definition of the partial derivative. By Taylor's Theorem we have

$$g(x) = \sum_{|\beta|\leq s} \frac{\partial^\beta g(0)}{\beta!} x^\beta + \tilde{\varphi}(x) \tag{2.40}$$

with $\tilde{\varphi}(x) = o(\|x\|^s)$. Since $g(x) = o(\|x\|^p)$, we have $\partial^\beta g(0) = 0$ for all $|\beta| \leq p$. Taylor's Theorem for $\partial^\alpha g(x)$ yields

$$\partial^\alpha g(x) = \sum_{|\beta|\leq s-|\alpha|} \frac{\partial^{\alpha+\beta} g(0)}{\beta!} x^\beta + \varphi^*(x) \tag{2.41}$$

with $\varphi^*(x) = o(\|x\|^{s-|\alpha|})$. Since we know that $\partial^{\alpha+\beta} g(0) = 0$ for $|\alpha + \beta| \leq p$,

$$\partial^\alpha g(x) = \sum_{p+1-|\alpha|\leq|\beta|\leq s-|\alpha|} \frac{\partial^{\alpha+\beta} g(0)}{\beta!} x^\beta + \varphi^*(x),$$

i.e. $\partial^\alpha g(x) = o(\|x\|^{p-|\alpha|})$.

The proof of 2. is similar, here we have $\partial^\beta g(0) = 0$ for all $|\beta| \le p-1$. □

Lemma 2.60. *Let $s, G, H \in \mathbb{N}$ with $s \ge G \ge H$. Let $g \in C^s(U, \mathbb{R}^n)$, where U is an open neighborhood of 0 in $\mathbb{R}^n$, and let $g(x) = o(\|x\|^G)$ for $x \to 0$. Let $h(x)$ be a polynomial with respect to $x \in \mathbb{R}^n$ and let $c_1, c_2 > 0$ be such that $c_1\|x\|^H \le h(x) \le c_2\|x\|^H$ holds for all $x \in U$.*

Then $\frac{g(x)}{h(x)} \in C^{G-H}(U, \mathbb{R}^n)$, where $\left(\partial^\alpha \frac{g}{h}\right)(x) = 0$ for all $0 \le |\alpha| \le G-H$.

Proof: $\frac{g(x)}{h(x)} \in C^s(U \setminus \{0\}, \mathbb{R}^n)$, since $h(x) \neq 0$ for $x \in U \setminus \{0\}$. We show that for $k = |\alpha| \le G-H$

$$\partial_x^\alpha \frac{g(x)}{h(x)} = \frac{g_\alpha(x)}{h_\alpha(x)} \tag{2.42}$$

holds for all $x \in U \setminus \{0\}$ with functions $g_\alpha(x) = o(\|x\|^{G-k+(2^k-1)H})$ for $x \to 0$ and $h_\alpha(x) = [h(x)]^{2^k}$, i.e. $c_1'\|x\|^{2^kH} \le h_\alpha(x) \le c_2'\|x\|^{2^kH}$. Then

$$\partial_x^\alpha \frac{g(x)}{h(x)} = o(\|x\|^{G-H-|\alpha|}). \tag{2.43}$$

We prove (2.42) by induction with respect to k. For $k=0$, (2.42) is true. Denoting $\partial_i := \frac{\partial}{\partial x_i}$, we have

$$\partial_i \frac{g_\alpha(x)}{h_\alpha(x)} = \frac{\partial_i g_\alpha(x) \cdot h_\alpha(x) - \partial_i h_\alpha(x) \cdot g_\alpha(x)}{h_\alpha(x)^2}$$

with $G - H - 1 \ge |\alpha| = k \ge 0$. With $e_i = (0, \ldots, 0, 1, 0, \ldots, 0)$, where the 1 is at i-th position, set $g_{\alpha+e_i}(x) := \partial_i g_\alpha(x) \cdot h_\alpha(x) - \partial_i h_\alpha(x) \cdot g_\alpha(x) = o(\|x\|^{G-k+(2^k-1)H+2^kH-1}) = o(\|x\|^{G-(k+1)+(2^{k+1}-1)H})$ by Lemma 2.59. Moreover, $h_{\alpha+e_i}(x) = h_\alpha(x)^2$. This shows (2.42) and thus (2.43).

Hence, the partial derivatives of order $|\alpha| \le G-H$ at the origin vanish. Indeed, for $\alpha = 0$ this is clear by (2.43). For $|\alpha| \ge 1$ set $\alpha = \alpha' + e_i$ with $|\alpha'| \le G-H-1$. We have

$$\begin{aligned}
\partial^{\alpha'+e_i} \frac{g(0)}{h(0)} &= \lim_{a\to 0} \frac{\partial^{\alpha'} \frac{g(ae_i)}{h(ae_i)}}{a} \\
&\le \lim_{a\to 0} \frac{o(|a|^{G-H-|\alpha'|})}{a} && \text{by (2.43)} \\
&= 0 && \text{since } G-H-|\alpha'| \ge 1,
\end{aligned}$$

where $a \in \mathbb{R}$ and $e_i \in \mathbb{R}^n$ is the i-th unit vector.

The partial derivatives are continuous since for $G - H - |\alpha| \ge 0$ we have $\lim_{x\to 0} \partial^\alpha \frac{g(x)}{h(x)} = \lim_{x\to 0} o(\|x\|^{G-H-|\alpha|}) = 0$ by (2.43). □

We prove 3. of Proposition 2.58: By Lemma 2.53 we have $V(x) = \mathfrak{n}(x) + \varphi(x)$ with $\varphi \in C^\sigma(A(x_0), \mathbb{R}^n)$ and $\varphi(x) = o(\|x - x_0\|^P)$. Hence, for $x \in A(x_0) \setminus \{x_0\}$

$$W(x) = \frac{V(x)}{\mathfrak{n}(x)} = 1 + \frac{\varphi(x)}{\mathfrak{n}(x)}.$$

Now set $g(x) := \varphi(x + x_0)$ and $h(x) := \mathfrak{n}(x + x_0)$ and use Lemma 2.60 and (2.39) where $G = P$ and $H = 2$. This proves 3. □

We will later calculate the Taylor polynomial to use the local information which is encoded in it. The function $W(x) = \frac{V(x)}{\mathfrak{n}(x)}$ satisfies a certain partial differential equation. We will later approximate the function W by w and obtain an approximation $v_W(x) = w(x)\mathfrak{n}(x)$ of $V(x)$.

Example 2.61 *We apply this procedure to the example (2.11) and calculate the following Taylor polynomial for V of order four, which even turns out to be of order five:*

$$\mathfrak{h}(x,y) = \frac{1}{2}x^2 + \frac{1}{2}y^2 + x^4 - \frac{3}{2}x^3y + \frac{11}{16}x^2y^2 + \frac{1}{32}xy^3 + \frac{3}{32}y^4.$$

In this case $\mathfrak{n}(x) = \mathfrak{h}(x) > 0$ holds for all $x \neq x_0$.

2.3.5 Summary and Examples

In Section 2.3 we have defined the global Lyapunov functions V and T. T satisfies the equation $T'(x) = -\bar{c}$ with a constant orbital derivative and thus is not defined in x_0. V fulfills $V'(x) = -p(x)$ where the function $p(x)$ satisfies the conditions of Theorem 2.46. In particular, $p(x_0) = 0$ and thus V is defined and smooth in x_0. T and V are both C^σ-functions, T is defined on $A(x_0) \setminus \{x_0\}$ and V is defined on $A(x_0)$. T provides information about the time which a solution takes from one point to another.

For the approximation, the equation $T'(x) = -\bar{c}$ with a constant orbital derivative is preferable. Moreover, the function $p(x)$ requires knowledge about the position of x_0, whereas the constant $\bar{c}$ does not. The main disadvantage of T is the fact, that T is not defined in x_0.

The function V is unique up to a constant. This relies on the smoothness of V in x_0. The values of T, however, can be fixed arbitrarily on a non-characteristic hypersurface. If we do the same with a function V^* satisfying $(V^*)'(x) = -p(x)$ for all $x \in A(x_0) \setminus \{x_0\}$, i.e. we assume that $V^*(x) = H(x)$ holds for all $x \in \Omega$, where Ω is a non-characteristic hypersurface, then V^* cannot be continued in a continuous way to x_0 in general and is only defined for $x \in A(x_0) \setminus \{x_0\}$, similarly to T.

At the end of this section we give simple examples where we can explicitly calculate all different Lyapunov functions.

Example 2.62 *We consider the* linear *scalar differential equation*

$$\dot{x} = ax$$

with $x \in \mathbb{R}$ *and* $a = -\nu < 0$. $x_0 = 0$ *is an exponentially asymptotically stable equilibrium. As a non-characteristic hypersurface we choose* $\Omega = \{-1, 1\}$. *With the function* $H\colon \Omega \to \mathbb{R}$ *defined by* $H(-1) = c_-$, $H(1) = c_+$ *we obtain the function*

$$T(x) = \bar{c}\,\frac{\ln|x|}{\nu} + c_{sgn(x)} \text{ for } x \neq 0.$$

Indeed, following the proof of Theorem 2.38 we determine $\theta(x)$ *by* $S_{\theta(x)}x \in \Omega$. *Since* $S_t x = xe^{at}$, $|S_\theta x| = 1$ *if and only if* $\theta = \frac{1}{a}\ln\frac{1}{|x|}$. *Now set* $T(x) = \bar{c}\,\theta(x) + H(S_{\theta(x)}x)$.

The functions $\mathfrak{v}$ *and* V *with* $p(x) = x^2$ *are given by* $V(x) = \mathfrak{v}(x) = \frac{1}{2\nu}x^2$, *since the equation is linear, cf. Corollary 2.50. The function* $\mathfrak{d}$ *is not unique but only defined up to a constant, and is thus always* $\mathfrak{d}(x) = cx^2$ *with* $c > 0$ *in the scalar case.*

Example 2.63 *Consider the* nonlinear *scalar differential equation*

$$\dot{x} = -x - x^3.$$

$x_0 = 0$ *is an asymptotically stable equilibrium. We calculate the different functions explicitly, cf. also Figure 2.2, where* $p(x) = x^2$, $\Omega = \{-1, 1\}$ *and* $H\colon \Omega \to \mathbb{R}$ *is defined by* $H(-1) = H(1) = 0$. *Then*

$$\begin{aligned}
\mathfrak{v}(x) &= \frac{1}{2}x^2\\
V(x) &= \frac{1}{2}\ln(1+x^2)\\
T(x) &= \frac{\bar{c}}{2}\ln\frac{2x^2}{1+x^2}\\
L(x) &= \left(\frac{2x^2}{1+x^2}\right)^{\frac{\bar{c}}{2}},
\end{aligned}$$

For the function $\mathfrak{d}$ *we have* $\mathfrak{d}(x) = cx^2$ *with any* $c > 0$, *cf. the preceding example.*

PROOF: The solution $x(t)$ of the differential equation with initial value $x(0) = \xi$ is calculated by separation and is given by

$$x(t) = \frac{\frac{\xi}{\sqrt{1+\xi^2}}e^{-t}}{\sqrt{1 - \frac{\xi^2}{1+\xi^2}e^{-2t}}} \text{ for } T \geq 0.$$

We calculate $V(x)$ satisfying $V'(x) = -x^2$ and substitute $y = \frac{x}{\sqrt{1+x^2}}e^{-t}$. Note that $|y| < 1$ for $t \geq 0$.

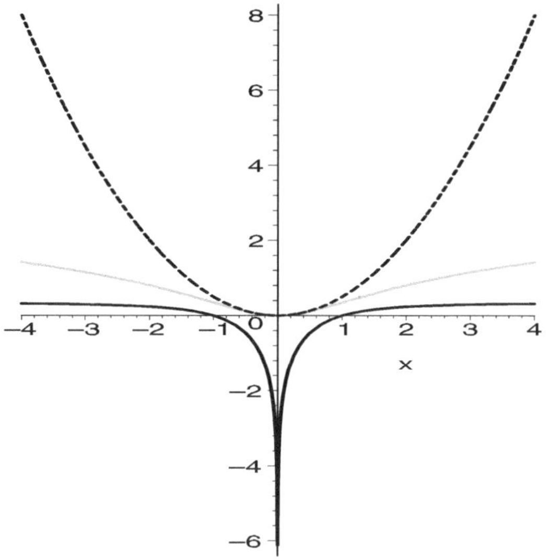

Fig. 2.2. The functions $\mathfrak{v}(x)$ (dotted), $V(x)$ (grey) and $T(x)$ (black solid) with $\bar{c} = 1$, cf. Example 2.63. Note that $\mathfrak{v}$ is the Taylor polynomial of V of order two and that $T(x)$ tends to $-\infty$ as x approaches the equilibrium $x_0 = 0$.

$$
\begin{aligned}
V(x) &= \int_0^\infty (S_t x)^2\, dt = \int_0^\infty \frac{\frac{x^2}{1+x^2} e^{-2t}}{1 - \frac{x^2}{1+x^2} e^{-2t}}\, dt = \int_{\frac{x}{\sqrt{1+x^2}}}^0 \frac{y^2}{1-y^2} \frac{dy}{-y} \\
&= \int_{\frac{x}{\sqrt{1+x^2}}}^0 \frac{-y}{1-y^2}\, dy = \frac{1}{2} \ln(1-y^2) \Big|_{\frac{x}{\sqrt{1+x^2}}}^0 = \frac{1}{2} \ln(1+x^2).
\end{aligned}
$$

We calculate the function θ by $|S_{\theta(x)} x|^2 = 1$ for $x \neq 0$. Hence

$$
\begin{aligned}
\frac{x^2}{1+x^2} e^{-2\theta} &= 1 - \frac{x^2}{1+x^2} e^{-2\theta} \\
e^{-2\theta} &= \frac{1+x^2}{2x^2} \\
\theta(x) &= \frac{1}{2} \ln \frac{2x^2}{1+x^2}.
\end{aligned}
$$

The function T satisfying $T'(x) = -\bar{c}$ and $T(\pm 1) = H(\pm 1) = 0$ is then given by $T(x) = \bar{c}\,\theta(x) = \frac{\bar{c}}{2} \ln \frac{2x^2}{1+x^2}$.

Thus, the function $L(x) = e^{T(x)}$ is given by $L(x) = \left(\frac{2x^2}{1+x^2}\right)^{\frac{\bar{c}}{2}}$. The smoothness of L at 0 depends on $\bar{c}$: For $\bar{c} = 1$, L is continuous in zero but not differentiable since $\lim_{x \searrow 0} \frac{L(x)-L(0)}{x} = \lim_{x \searrow 0} \sqrt{\frac{2}{1+x^2}} = \sqrt{2}$ but $\lim_{x \nearrow 0} \frac{L(x)-L(0)}{x} = \lim_{x \searrow 0} \left(-\sqrt{\frac{2}{1+x^2}}\right) = -\sqrt{2}$. □

3

Radial Basis Functions

In order to construct a function with negative orbital derivative, we approximate the Lyapunov function $Q = V$ or $Q = T$ which satisfies the linear partial differential equation $V'(x) = -p(x)$, $T'(x) = -\bar{c}$, respectively, by *radial basis functions*. Radial basis functions are a powerful tool to approximate multivariate functions or to solve partial differential equations. For an overview cf. [63], [11], [10], or [52], for a tutorial cf. [40]. The main advantage of this method is that it is meshless, i.e. no triangulation of the space $\mathbb{R}^n$ is needed.

We will first discuss how radial basis functions are used to interpolate a function by given values. This is used if the function values are only known on a grid and one is interested in reconstructing the original function. We will later indicate how we use this part of the radial basis functions theory in our setting. To interpolate a function Q by given values on a grid $\{\xi_1, \ldots, \xi_M\}$, one uses an ansatz of a linear combination of shifted radial basis functions centered in each grid point, i.e. $q(x) = \sum_{k=1}^{M} \gamma_k \Psi(x - \xi_k)$ where Ψ is a fixed function which often is of the form $\Psi(x) = \psi(\|x\|)$ and hence is called radial basis function. The coefficients γ_k of the linear combination are determined by the claim that the values of the interpolant q and the original function Q are the same for all points of the grid, i.e. $q(\xi_j) = Q(\xi_j)$ for all $j = 1, \ldots, M$. This condition is equivalent to a system of linear equations for γ; its size is the number of the grid points M. Error estimates of $|Q(x) - q(x)|$, i.e. the difference of the approximated function Q to the approximation q, can be derived and depend on the density of the grid.

A generalization, however, enables us to interpolate functions via the values of a *linear operator* applied to the function, in our case via their orbital derivative. In other words, we approximate the solution of a linear partial differential equation. A combination of both, orbital derivatives and function values, is used to approximate solutions of a Cauchy problem in partial differential equations.

Let us explain the approximation with radial basis functions using a linear operator D, e.g. the orbital derivative $DQ(x) = Q'(x) = \langle \nabla Q(x), f(x) \rangle$. We use the symmetric ansatz leading to a symmetric interpolation matrix A.

One defines a grid $\{x_1, \ldots, x_N\}$. The reconstruction (approximation) of a function Q through the values of Q' on the grid is obtained by the ansatz $q(x) = \sum_{k=1}^N \beta_k \langle \nabla_y \Psi(x-y)\big|_{y=x_k}, f(x_k)\rangle$. The function Ψ is again the radial basis function. The coefficients $\beta_k \in \mathbb{R}$ are determined by the claim that $q'(x_j) = Q'(x_j)$ holds for all grid points $j = 1, \ldots, N$. This is equivalent to a system of linear equations $A\beta = \alpha$ where the interpolation matrix A and the right-hand side vector α are determined by the grid and the values $Q'(x_j)$. The interpolation matrix A is an $(N \times N)$ matrix, where N is the number of grid points. We show that A is positive definite and thus the linear equation has a unique solution β. Provided that Q is smooth enough, one obtains an error estimate on $|Q'(x) - q'(x)|$ depending on the density of the grid.

While the interpolation of function values has been studied in detail since the 1970's, the interpolation via the values of a linear operator has only been considered since the 1990's. The values of such linear operators are also called Hermite-Birkhoff data. They have been studied, e.g. by Iske [38], Wu [67], Wendland [63] and Franke & Schaback [17] and [18]. The last authors approximate the solution of a Cauchy problem in partial differential equations, cf. also [54]. This results in a mixed problem, combining different linear operators, cf. [17] and [18]. Their error estimates, however, use the fact that the linear operator is translation invariant and, hence, they study partial differential equations with constant coefficients. Our linear operator D, however, is the orbital derivative and is not translation invariant. Error estimates hence have to use different techniques, which are Taylor expansions in our case.

Since in our case the values of $Q'(x)$ are known for all points, we are free to choose any grid, provided that the grid points are pairwise distinct and no equilibria. A denser grid provides a smaller error but on the other hand the condition number of the interpolation matrix becomes larger. Schaback observed this dilemma and called it the *uncertainty principle* for radial basis function interpolation, cf. [53]. Iske, [39] or [40], showed that the optimal choice to obtain a dense grid and, at the same time, a large minimal distance of grid points which corresponds to a small condition number of the linear equation, is the hexagonal grid and its generalizations to higher dimensions. Thus, we use such a grid in general, however, we can add additional points where the grid appears not to be dense enough.

We are interested in the determination of the basin of attraction of an equilibrium x_0. As we have seen in the last chapter, the key is to construct a Lyapunov function q, i.e. a function with negative orbital derivative in a set $K \setminus \{x_0\}$ with $x_0 \in \overset{\circ}{K}$. Then there exists a sublevel set of q in K by Theorem 2.45, which is a Lyapunov basin and thus a subset of the basin of attraction by Theorem 2.24.

Since we are not able to determine a Lyapunov function explicitly in general, we will approximate a Lyapunov function Q by the approximation q using radial basis functions. In Chapter 2 we have proved the existence of the Lyapunov functions $Q = T$ and $Q = V$ which have

- known orbital derivatives in the basin of attraction
- and, in the case of T, known values at a non-characteristic hypersurface.

In other words, V satisfies a linear partial differential equation and T a Cauchy problem. We have shown that these problems have unique and smooth solutions. Using radial basis functions, we are thus able to approximate V, T, respectively. It turns out that the approximation itself has negative orbital derivative and thus is a Lyapunov function.

In this chapter we will present the facts from radial basis function theory that are needed for our goals. We consider a general function Q which later will be one of the Lyapunov functions T or V of Chapter 2. More details of this application to Lyapunov functions will be studied in Chapters 4 and 5.

The approximation is explained in Section 3.1, whereas the error estimates are given in Section 3.2. As radial basis functions we use the Wendland functions, which have compact support and are polynomials on their support.

3.1 Approximation

Radial basis functions are a powerful tool to approximate multivariate functions. The basic idea is to interpolate a function Q by given values on a grid. One uses an ansatz of a linear combination of shifted radial basis functions centered in each grid point. The shifted radial basis function is thus a function of the distance to the respective grid point. The coefficients of the linear combination are determined by the claim that the values of the interpolant q and the original function Q are the same for all points of the grid. One obtains the coefficients by solving a system of linear equations; its size is determined by the number of the grid points. In Section 3.1.1 the values of the function and in Section 3.1.2 the values of a general linear operator, in particular the orbital derivative, are given on a grid. Section 3.1.3 considers the case where the values of two different linear operators, in particular the functions values and the values of the orbital derivatives, are prescribed; this is called mixed approximation. In Section 3.1.4 we present the Wendland functions, the family of radial basis functions which is used in this book.

3.1.1 Approximation via Function Values

Consider a function $Q\colon \mathbb{R}^n \to \mathbb{R}$. Assume, the values $Q(\xi_k)$ are given for all points $\xi_k \in X_M^0$, where $X_M^0 = \{\xi_1, \ldots, \xi_M\} \subset \mathbb{R}^n$ is a set of pairwise distinct points. For the approximant $q\colon \mathbb{R}^n \to \mathbb{R}$ we make the following ansatz:

$$q(x) = \sum_{k=1}^{M} \gamma_k \Psi(x - \xi_k), \tag{3.1}$$

where $\Psi(x) := \psi(\|x\|)$ is a fixed function, the radial basis function. Note that many results also hold for functions Ψ which are not of this special form. We, however, assume in the following that Ψ is a function of $\|x\|$.

The coefficients $\gamma_k \in \mathbb{R}$ are determined by the claim that

$$Q(\xi_j) = q(\xi_j) \tag{3.2}$$

holds for all $\xi_j \in X_M^0$. Plugging the ansatz (3.1) into (3.2) one obtains

$$Q(\xi_j) = \sum_{k=1}^{M} \gamma_k \Psi(\xi_j - \xi_k)$$

for all $j = 1, \ldots, M$. This is equivalent to the following system of linear equations with interpolation matrix $A^0 = (a^0_{jk})_{j,k=1,\ldots,M}$ and vector $\alpha^0 = (\alpha^0_j)_{j=1,\ldots,M}$

$$\begin{aligned} A^0\gamma &= \alpha^0, \\ \text{where } a^0_{jk} &= \Psi(\xi_j - \xi_k) \\ \text{and } \alpha^0_j &= Q(\xi_j). \end{aligned}$$

Note that the interpolation matrix A^0 is symmetric by construction.

Definition 3.1 (Interpolation problem, function values). *Let X_M^0 be a grid with pairwise distinct points and $Q\colon \mathbb{R}^n \to \mathbb{R}$.*

The interpolation matrix $A^0 = (a^0_{jk})_{j,k=1,\ldots,M}$ *is given by*

$$a^0_{jk} = \Psi(\xi_j - \xi_k).$$

The reconstruction q of Q with respect to the grid X_M^0 *is given by*

$$q(x) = \sum_{k=1}^{M} \gamma_k \Psi(x - \xi_k),$$

where γ is the solution of $A^0\gamma = \alpha^0$ with $\alpha^0_j = Q(\xi_j)$ and $\Psi \in C^0(\mathbb{R}^n, \mathbb{R})$ is a radial basis function.

For existence and uniqueness of the solution γ, the interpolation matrix A^0 must have full rank. We will even show that A^0 is positive definite. Note that the matrix A^0 depends on the grid X_M^0 and on Ψ. A function Ψ for which the interpolation matrix A^0 is positive definite for all grids X_M^0 is called a positive definite function.

Definition 3.2 (Positive definite function). *A function $\Psi \in C^0(\mathbb{R}^n, \mathbb{R})$ is called* positive definite, *if for all $M \in \mathbb{N}$ and all grids $X_M^0 = \{\xi_1, \ldots, \xi_M\}$ with pairwise distinct points ξ_k*

$$\gamma^T A^0 \gamma = \sum_{j,k=1}^{M} \gamma_j \gamma_k \Psi(\xi_j - \xi_k) > 0 \tag{3.3}$$

holds for all $\gamma = (\gamma_1, \ldots, \gamma_M) \in \mathbb{R}^M \setminus \{0\}$, where A^0 is defined as in Definition 3.1.

More generally, one defines conditionally positive definite functions of a certain degree, for which (3.3) only holds for all γ satisfying additional conditions. On the other hand the ansatz for q then includes polynomials of a certain degree in order to achieve uniqueness of the solution. Many radial basis functions such as thin plate splines and Multiquadrics are only conditionally positive definite. Since the Wendland functions, which will be used in this book, are positive definite functions, we restrict ourselves to positive definite functions in the following.

Definition 3.2 implies the following lemma.

Lemma 3.3. *Let $\Psi(x) = \psi(\|x\|)$ be a positive definite function.*
Then for all grids X_M^0 with pairwise distinct points and all functions Q the corresponding interpolation problem, cf. Definition 3.1, has a unique solution. The interpolation matrix A^0 is symmetric and positive definite.

Since A^0 is symmetric and positive definite, one can use the Cholesky-method, cf. for instance [55], p. 37, to solve the system of linear equations. When using a denser grid one can thus use the information of the points calculated before.

A criterion for a function Ψ to be positive definite using its Fourier transform will be given in Section 3.2.2. The Wendland functions are positive definite functions; they will be introduced in Section 3.1.4.

3.1.2 Approximation via Orbital Derivatives

In this section we use a generalization of the radial basis function theory, cf. [38], [17], [63]: The approximating function is not obtained by interpolation of the function values, but by interpolation of the values of a linear operator applied to the function. The interpolation of function values discussed in Section 3.1.1 is a special case where the linear operator is the identity, i.e. $D = \text{id}$. In this section we study a general linear operator D. In two subsections we later calculate the explicit formulas for D being the orbital derivative, i.e. $DQ(x) = Q'(x) = \langle \nabla Q(x), f(x) \rangle$, and for the operator D_m of the orbital derivative and multiplication with the scalar-valued function $m(x)$, namely $D_m Q(x) = Q'(x) + m(x)Q(x)$.

Consider a function $Q\colon \mathbb{R}^n \to \mathbb{R}$, which is smooth enough such that DQ is defined. For example, if Q is the orbital derivative then we claim that $Q \in C^1(\mathbb{R}^n, \mathbb{R})$. Assume that the values $(\delta_{x_k} \circ D)^x Q(x) = (DQ)(x_k)$ are known for all points $x_k \in X_N$, where $X_N = \{x_1, \ldots, x_N\}$ is a set of pairwise distinct points. Here, δ_{x_k} denotes Dirac's δ-operator, i.e. $\delta_{x_k} Q(x) = Q(x_k)$ and the superscript x denotes the application of the operator with respect to the variable x. For the approximant $q\colon \mathbb{R}^n \to \mathbb{R}$ we make the following ansatz:

$$q(x) = \sum_{k=1}^{N} \beta_k (\delta_{x_k} \circ D)^y \Psi(x - y), \tag{3.4}$$

where $\Psi(x) := \psi(\|x\|)$ is a fixed function, the radial basis function, and $\beta_k \in \mathbb{R}$. Note that this ansatz requires a certain smoothness of the function Ψ if the operator D is a differential operator. The advantage of this ansatz including the operator D is, as it turns out, that the interpolation matrix A will be symmetric. Note that for $D = \text{id}$ the ansatz (3.4) is (3.1). Using a different ansatz than (3.4) which does not include the operator D, results in the unsymmetric approach, cf. Kansa [43] and [44]. We, however, will concentrate on the symmetric approach in this book.

The coefficients β_k are determined by the claim that

$$(\delta_{x_j} \circ D)^x Q(x) = (\delta_{x_j} \circ D)^x q(x) \tag{3.5}$$

holds for all $x_j \in X_N$, i.e. $DQ(x_j) = Dq(x_j)$. Plugging the ansatz (3.4) into (3.5) one obtains

$$\begin{aligned}(\delta_{x_j} \circ D)^x Q(x) &= (\delta_{x_j} \circ D)^x q(x)\\ &= \sum_{k=1}^{N} (\delta_{x_j} \circ D)^x (\delta_{x_k} \circ D)^y \Psi(x-y)\beta_k\end{aligned}$$

for all $j = 1, \ldots, N$, since D is a linear operator. This is equivalent to the following system of linear equations with interpolation matrix $A = (a_{jk})_{j,k=1,\ldots,N}$ and vector $\alpha = (\alpha_j)_{j=1,\ldots,N}$

$$\begin{aligned}A\beta &= \alpha,\\ \text{where } a_{jk} &= (\delta_{x_j} \circ D)^x (\delta_{x_k} \circ D)^y \Psi(x-y)\\ \text{and } \alpha_j &= (\delta_{x_j} \circ D)^x Q(x).\end{aligned}$$

A clearly is a symmetric matrix. Note that Ψ has to be smooth enough to apply the operator D twice to Ψ. If, for example, D is a differential operator of first order as the orbital derivative, Ψ has to be at least C^2. We summarize the interpolation problem for a general linear operator D.

Definition 3.4 (Interpolation problem, operator). *Let X_N be a grid with pairwise distinct points and $Q\colon \mathbb{R}^n \to \mathbb{R}$. Let D be a linear operator.*

The interpolation matrix $A = (a_{jk})_{j,k=1,\ldots,N}$ is given by

$$a_{jk} = (\delta_{x_j} \circ D)^x (\delta_{x_k} \circ D)^y \Psi(x-y). \tag{3.6}$$

The reconstruction q of Q with respect to the grid X_N and the operator D *is given by*

$$q(x) = \sum_{k=1}^{N} \beta_k (\delta_{x_k} \circ D)^y \Psi(x-y),$$

where β is the solution of $A\beta = \alpha$ with $\alpha_j = (\delta_{x_j} \circ D)^x Q(x)$. Note that Q and Ψ have to be smooth enough in order to apply D.

For existence and uniqueness of the solution β, the interpolation matrix A must have full rank. We will even obtain a positive definite matrix A for grids which include no equilibrium point. A criterion for the function Ψ to guarantee that all interpolation matrices are positive definite will be given in Section 3.2.2. The Wendland functions are functions with which one obtains positive definite interpolation matrices.

We will now consider the operator D of the orbital derivative in more detail. Later, we consider the linear operator D_m, defined by $D_mQ(x) = Q'(x) + m(x) \cdot Q(x)$ which combines orbital derivative and multiplication with a scalar-valued function m.

Orbital Derivative

We consider the autonomous differential equation $\dot{x} = f(x)$, $x \in \mathbb{R}^n$. Let x_0 be an equilibrium point of the corresponding dynamical system.

We consider the linear operator of the orbital derivative

$$DQ(x) := \langle \nabla Q(x), f(x) \rangle = Q'(x).$$

Let $X_N = \{x_1, \ldots, x_N\} \subset \mathbb{R}^n$ be a set of pairwise distinct points, which are no equilibrium points, i.e. $f(x_j) \neq 0$ for all $j = 1, \ldots, N$. If x_j was an equilibrium point, then we would have $a_{jk} = 0$ for all $k = 1, \ldots, N$ since $f(x_j) = 0$, cf. (3.6) and the definition of D. Hence, A would be a singular matrix. Later we will see that for suitable radial basis functions, e.g. the Wendland functions, excluding equilibrium points is not only a necessary but also a sufficient condition for the positive definiteness of the interpolation matrix, cf. Proposition 3.23.

Let $\Psi(x) := \psi(\|x\|)$ be a suitable radial basis function with $\Psi \in C^2(\mathbb{R}^n, \mathbb{R})$. We calculate the matrix elements a_{jk} of A defined in Definition 3.4.

Proposition 3.5 *Let D be given by*

$$DQ(x) = \langle \nabla Q(x), f(x) \rangle.$$

Let $\Psi(x) := \psi(\|x\|)$ with $\Psi \in C^2(\mathbb{R}^n, \mathbb{R})$. Define ψ_1 and ψ_2 by

$$\psi_1(r) = \frac{\frac{d}{dr}\psi(r)}{r} \text{ for } r > 0, \tag{3.7}$$

$$\psi_2(r) = \begin{cases} \frac{\frac{d}{dr}\psi_1(r)}{r} & \text{for } r > 0 \\ 0 & \text{for } r = 0 \end{cases}. \tag{3.8}$$

Let ψ be such that $\frac{d}{dr}\psi(r) = O(r)$, ψ_1 can be extended continuously to 0 and $\frac{d}{dr}\psi_1(r) = O(1)$ for $r \to 0$.

The matrix elements a_{jk} of the interpolation matrix A in Definition 3.4 are then given by

$$\begin{aligned} a_{jk} = &\ \psi_2(\|x_j - x_k\|)\langle x_j - x_k, f(x_j)\rangle\langle x_k - x_j, f(x_k)\rangle \\ &-\psi_1(\|x_j - x_k\|)\langle f(x_j), f(x_k)\rangle. \end{aligned} \tag{3.9}$$

The approximant q and its orbital derivative are given by

$$q(x) = \sum_{k=1}^{N} \beta_k \langle x_k - x, f(x_k)\rangle \psi_1(\|x - x_k\|),$$

$$q'(x) = \sum_{k=1}^{N} \beta_k \Big[\psi_2(\|x - x_k\|)\langle x - x_k, f(x)\rangle\langle x_k - x, f(x_k)\rangle$$
$$-\psi_1(\|x - x_k\|)\langle f(x), f(x_k)\rangle\Big].$$

Proof: The assumptions on ψ_1 imply that $\psi_2(r) = O\left(\frac{1}{r}\right)$ holds for $r \to 0$. Thus, for the first term of (3.9) we have $\psi_2(\|x_j - x_k\|)\langle x_j - x_k, f(x_j)\rangle\langle x_k - x_j, f(x_k)\rangle = O(\|x_j - x_k\|)$ for $r = \|x_j - x_k\| \to 0$. Hence, the terms in (3.9) are well-defined and continuous for $\|x_j - x_k\| = r \in [0, \infty)$.

For the following calculation we denote $' = \frac{d}{dr}$. By (3.6) we have for $x_j \neq x_k$

$$\begin{aligned}
a_{jk} &= (\delta_{x_j} \circ D)^x (\delta_{x_k} \circ D)^y \Psi(x - y) \\
&= (\delta_{x_j} \circ D)^x \langle \nabla_y \psi(\|x - y\|)\Big|_{y=x_k}, f(x_k)\rangle \\
&= (\delta_{x_j} \circ D)^x \left[\frac{\psi'(\|x - x_k\|)}{\|x - x_k\|}\langle x_k - x, f(x_k)\rangle\right] \\
&= (\delta_{x_j} \circ D)^x \left[\psi_1(\|x - x_k\|)\langle x_k - x, f(x_k)\rangle\right] \\
&= \frac{\psi_1'(\|x - x_k\|)}{\|x - x_k\|}\langle x - x_k, f(x)\rangle\langle x_k - x, f(x_k)\rangle\Big|_{x=x_j} \\
&\quad -\psi_1(\|x_j - x_k\|)\langle f(x_j), f(x_k)\rangle \\
&= \psi_2(\|x_j - x_k\|)\langle x_j - x_k, f(x_j)\rangle\langle x_k - x_j, f(x_k)\rangle \\
&\quad -\psi_1(\|x_j - x_k\|)\langle f(x_j), f(x_k)\rangle.
\end{aligned}$$

For $x_j = x_k$ we have $a_{jj} = -\psi_1(0)\|f(x_j)\|^2$. The formulas for $q(x)$ and $q'(x)$ follow by similar calculations. □

Orbital Derivative and Multiplication

We consider again the autonomous differential equation $\dot{x} = f(x)$ as above with equilibrium x_0. We fix a function $m\colon \mathbb{R}^n \setminus \{x_0\} \to \mathbb{R}$ which is continuous in $\mathbb{R}^n \setminus \{x_0\}$ and bounded for $x \to x_0$. m will later be the function defined in Proposition 3.38. We consider the linear operator D_m defined by $D_m Q(x) := Q'(x) + m(x)Q(x)$. We calculate the matrix elements of A defined in Definition 3.4 for the operator D_m.

Proposition 3.6 *Let $X_N = \{x_1, \ldots, x_N\}$ be a set of pairwise distinct points, which are no equilibrium points. Let $m\colon \mathbb{R}^n \setminus \{x_0\} \to \mathbb{R}$ be continuous in $\mathbb{R}^n \setminus \{x_0\}$ and bounded for $x \to x_0$. Let $D = D_m$ be given by*

$$D_m Q(x) := \langle \nabla Q(x), f(x)\rangle + m(x)Q(x) = Q'(x) + m(x)Q(x).$$

Let $\Psi(x) := \psi(\|x\|)$ with $\Psi \in C^2(\mathbb{R}^n, \mathbb{R})$. Moreover, let ψ satisfy the assumptions of Proposition 3.5, where ψ_1 and ψ_2 are defined by (3.7) and (3.8).

The matrix elements a_{jk} of the interpolation matrix A in Definition 3.4 are then given by

$$\begin{aligned} a_{jk} = & \psi_2(\|x_j - x_k\|)\langle x_j - x_k, f(x_j)\rangle\langle x_k - x_j, f(x_k)\rangle \\ & -\psi_1(\|x_j - x_k\|)\langle f(x_j), f(x_k)\rangle \\ & +m(x_k)\psi_1(\|x_j - x_k\|)\langle x_j - x_k, f(x_j)\rangle \\ & +m(x_j)\psi_1(\|x_j - x_k\|)\langle x_k - x_j, f(x_k)\rangle \\ & +m(x_j)m(x_k)\psi(\|x_j - x_k\|). \end{aligned} \tag{3.10}$$

The approximant q and its orbital derivative q' are given by

$$q(x) = \sum_{k=1}^{N} \beta_k \left[\langle x_k - x, f(x_k)\rangle \psi_1(\|x - x_k\|) + m(x_k)\psi(\|x - x_k\|)\right],$$

$$\begin{aligned} q'(x) = \sum_{k=1}^{N} \beta_k \Big[& \psi_2(\|x - x_k\|)\langle x - x_k, f(x)\rangle\langle x_k - x, f(x_k)\rangle \\ & -\psi_1(\|x - x_k\|)\langle f(x), f(x_k)\rangle \\ & +m(x_k)\psi_1(\|x - x_k\|)\langle x - x_k, f(x)\rangle \Big]. \end{aligned}$$

PROOF: The terms in (3.10) are well-defined and continuous with respect to $r = \|x_j - x_k\| \in [0, \infty)$ for $x_j \neq x_0$ and $x_k \neq x_0$, cf. the proof of Proposition 3.5. By (3.6) we have

$$\begin{aligned} a_{jk} &= (\delta_{x_j} \circ D_m)^x (\delta_{x_k} \circ D_m)^y \psi(\|x - y\|) \\ &= (\delta_{x_j} \circ D_m)^x \left(\psi_1(\|x - x_k\|)\langle x_k - x, f(x_k)\rangle + m(x_k)\psi(\|x - x_k\|)\right) \\ &= \frac{\psi_1'(\|x - x_k\|)}{\|x - x_k\|}\langle x - x_k, f(x)\rangle\langle x_k - x, f(x_k)\rangle\Big|_{x=x_j} \\ &\quad -\psi_1(\|x_j - x_k\|)\langle f(x_j), f(x_k)\rangle \\ &\quad +m(x_k)\frac{\psi'(\|x - x_k\|)}{\|x - x_k\|}\langle x - x_k, f(x)\rangle\Big|_{x=x_j} \\ &\quad +m(x_j)\psi_1(\|x_j - x_k\|)\langle x_k - x_j, f(x_k)\rangle + m(x_j)m(x_k)\psi(\|x_j - x_k\|). \end{aligned}$$

The formulas for $q(x)$ and $q'(x)$ follow by similar calculations. □

3.1.3 Mixed Approximation

We combine the Sections 3.1.1 and 3.1.2 to a mixed approximation. Because of the linearity of the ansatz, one can consider two or more different operators.

With this ansatz one can, for instance, approximate solutions of a Cauchy problem for partial differential equations, cf. [17].

We restrict ourselves to the case of the two linear operators $D^0 = \mathrm{id}$ and D. D will later be the orbital derivative. Hence, we assume that for the function Q

- the function values $Q(x)$ are known on the grid $x \in X_M^0$ and
- the values of $DQ(x)$ are known on the grid $x \in X_N$.

This information will be used to reconstruct Q by an approximant q, which interpolates the above data.

We choose two grids, $X_N := \{x_1, \ldots, x_N\}$ and $X_M^0 := \{\xi_1, \ldots, \xi_M\}$. The mixed ansatz for the approximant q of the function Q is

$$q(x) = \sum_{k=1}^{N} \beta_k (\delta_{x_k} \circ D)^y \Psi(x-y) + \sum_{k=1}^{M} \gamma_k (\delta_{\xi_k} \circ D^0)^y \Psi(x-y), \qquad (3.11)$$

where $D^0 = \mathrm{id}$. The coefficients $\beta_k, \gamma_k \in \mathbb{R}$ are determined by the claim that

$$(\delta_{x_j} \circ D)^x Q(x) = (\delta_{x_j} \circ D)^x q(x) \qquad (3.12)$$

$$(\delta_{\xi_i} \circ D^0)^x Q(x) = (\delta_{\xi_i} \circ D^0)^x q(x) \qquad (3.13)$$

holds for all $x_j \in X_N$ and all $\xi_i \in X_M^0$. Plugging the ansatz (3.11) into both (3.12) and (3.13), one obtains

$$\begin{aligned}(\delta_{x_j} \circ D)^x Q(x) &= \sum_{k=1}^{N} \beta_k (\delta_{x_j} \circ D)^x (\delta_{x_k} \circ D)^y \Psi(x-y) \\ &\quad + \sum_{k=1}^{M} \gamma_k (\delta_{x_j} \circ D)^x (\delta_{\xi_k} \circ D^0)^y \Psi(x-y) \\ \text{and } (\delta_{\xi_i} \circ D^0)^x Q(x) &= \sum_{k=1}^{N} \beta_k (\delta_{\xi_i} \circ D^0)^x (\delta_{x_k} \circ D)^y \Psi(x-y) \\ &\quad + \sum_{k=1}^{M} \gamma_k (\delta_{\xi_i} \circ D^0)^x (\delta_{\xi_k} \circ D^0)^y \Psi(x-y)\end{aligned}$$

for all $j = 1, \ldots, N$ and $i = 1, \ldots, M$. This is equivalent to the following system of linear equations $\begin{pmatrix} A & C \\ C^T & A^0 \end{pmatrix} \begin{pmatrix} \beta \\ \gamma \end{pmatrix} = \begin{pmatrix} \alpha \\ \alpha^0 \end{pmatrix}$. For the definition of the matrices, cf. the following definition, where we summarize the mixed interpolation problem.

Definition 3.7 (Mixed interpolation problem). *Let $X_N = \{x_1, \ldots, x_N\}$ and $X_M^0 = \{\xi_1, \ldots, \xi_M\}$ be grids with pairwise distinct points, respectively, and $Q\colon \mathbb{R}^n \to \mathbb{R}$. Let D and $D^0 = id$ be linear operators.*

The interpolation matrix $\begin{pmatrix} A & C \\ C^T & A^0 \end{pmatrix}$ *has the matrices* $A = (a_{jk})_{j,k=1,\ldots,N}$, $C = (c_{jk})_{j=1,\ldots,N,\, k=1,\ldots,M}$ *and* $A^0 = (a^0_{jk})_{j,k=1,\ldots,M}$ *with elements*

$$\begin{aligned} a_{jk} &= (\delta_{x_j} \circ D)^x (\delta_{x_k} \circ D)^y \Psi(x-y) \\ c_{jk} &= (\delta_{x_j} \circ D)^x (\delta_{\xi_k} \circ D^0)^y \Psi(x-y) \\ a^0_{jk} &= (\delta_{\xi_j} \circ D^0)^x (\delta_{\xi_k} \circ D^0)^y \Psi(x-y). \end{aligned}$$

The reconstruction q of Q with respect to the grids X_N and X^0_M and the operators D and D^0 *is given by*

$$q(x) = \sum_{k=1}^{N} \beta_k (\delta_{x_k} \circ D)^y \Psi(x-y) + \sum_{k=1}^{M} \gamma_k (\delta_{\xi_k} \circ D^0)^y \Psi(x-y),$$

where (β, γ) *is the solution of* $\begin{pmatrix} A & C \\ C^T & A^0 \end{pmatrix} \begin{pmatrix} \beta \\ \gamma \end{pmatrix} = \begin{pmatrix} \alpha \\ \alpha^0 \end{pmatrix}$. *The vectors* α *and* α^0 *are given by* $\alpha_j = (\delta_{x_j} \circ D)^x Q(x)$ *and* $\alpha^0_i = (\delta_{\xi_i} \circ D^0)^x Q(x)$.

Under certain assumptions we will obtain a positive definite matrix $\begin{pmatrix} A & C \\ C^T & A^0 \end{pmatrix}$ if the grid X_N includes no equilibrium point.

We consider the autonomous differential equation $\dot{x} = f(x)$. Let x_0 be an equilibrium point. We consider the linear operator of the orbital derivative $DQ(x) = Q'(x)$ and calculate the matrix elements of A, C and A^0 defined in Definition 3.7.

Proposition 3.8 *Let* D *be given by*

$$DQ(x) = Q'(x) = \langle \nabla Q(x), f(x) \rangle.$$

Let $\Psi(x) := \psi(\|x\|)$ *with* $\Psi \in C^2(\mathbb{R}^n, \mathbb{R})$. *Moreover, let* ψ *satisfy the assumptions of Proposition 3.5, where* ψ_1 *and* ψ_2 *are defined by (3.7) and (3.8).*

The matrix elements a_{jk}, c_{jk} *and* a^0_{jk} *of the interpolation matrix* $\begin{pmatrix} A & C \\ C^T & A^0 \end{pmatrix}$ *in Definition 3.7 are then given by*

$$\begin{aligned} a_{jk} &= \psi_2(\|x_j - x_k\|) \langle x_j - x_k, f(x_j) \rangle \langle x_k - x_j, f(x_k) \rangle \\ &\quad -\psi_1(\|x_j - x_k\|) \langle f(x_j), f(x_k) \rangle && (3.14) \\ c_{jk} &= \psi_1(\|x_j - \xi_k\|) \langle x_j - \xi_k, f(x_j) \rangle && (3.15) \\ a^0_{jk} &= \psi(\|\xi_j - \xi_k\|). && (3.16) \end{aligned}$$

The approximant q *and its orbital derivative are given by*

$$q(x) = \sum_{k=1}^{N} \beta_k \psi_1(\|x - x_k\|)\langle x_k - x, f(x_k)\rangle + \sum_{k=1}^{M} \gamma_k \psi(\|x - \xi_k\|),$$

$$q'(x) = \sum_{k=1}^{N} \beta_k \left[\psi_2(\|x - x_k\|)\langle x - x_k, f(x)\rangle\langle x_k - x, f(x_k)\rangle \right.$$
$$\left. -\psi_1(\|x - x_k\|)\langle f(x), f(x_k)\rangle\right]$$
$$+\sum_{k=1}^{M} \gamma_k \psi_1(\|x - \xi_k\|)\langle x - \xi_k, f(x)\rangle,$$

where $\begin{pmatrix}\beta\\ \gamma\end{pmatrix}$ *is the solution of* $\begin{pmatrix}A & C\\ C^T & A^0\end{pmatrix}\begin{pmatrix}\beta\\ \gamma\end{pmatrix} = \begin{pmatrix}\alpha\\ \alpha^0\end{pmatrix}$.
The vectors α *and* α^0 *are given by* $\alpha_j = Q'(x_j)$ *and* $\alpha_i^0 = Q(\xi_i)$.

PROOF: The matrix elements of A have already been calculated in Proposition 3.5, cf. (3.9). For the elements of C we have

$$\begin{aligned} c_{jk} &= (\delta_{x_j} \circ D)^x (\delta_{\xi_k} \circ D^0)^y \Psi(x - y) \\ &= \frac{\psi'(\|x_j - \xi_k\|)}{\|x_j - \xi_k\|}\langle x_j - \xi_k, f(x_j)\rangle \\ &= \psi_1(\|x_j - \xi_k\|)\langle x_j - \xi_k, f(x_j)\rangle. \end{aligned}$$

For A^0 we have, cf. also Definition 3.1,

$$a_{jk}^0 = (\delta_{\xi_j} \circ D^0)^x (\delta_{\xi_k} \circ D^0)^y \Psi(x - y) = \psi(\|\xi_j - \xi_k\|).$$

The formulas for the function q and its orbital derivative follow similarly. □

3.1.4 Wendland Functions

In this book we use a special class of radial basis functions: the Wendland functions, introduced by Wendland, cf. [61] or [62]. In contrast to many classical radial basis function, e.g. polynomials, thin plate splines or Multiquadrics, the Wendland functions are positive definite and not only *conditionally* positive definite. Hence, it is not necessary to modify the ansatz discussed above by adding polynomials of some degree. Moreover, in contrast to the Gaussians or the inverse Multiquadrics, which are positive definite radial basis functions, the Wendland functions have compact support and are polynomials on their support. Thus, one can approximate functions which are not necessarily C^∞ and the resulting interpolation matrix A of the linear equation $A\beta = \alpha$ is sparse.

We prove in Section 3.2.2 that the interpolation matrix A is positive definite and, hence, the above linear equation $A\beta = \alpha$ has a unique solution β. For the proof of this fact as well as for the error estimates we use the *native space*

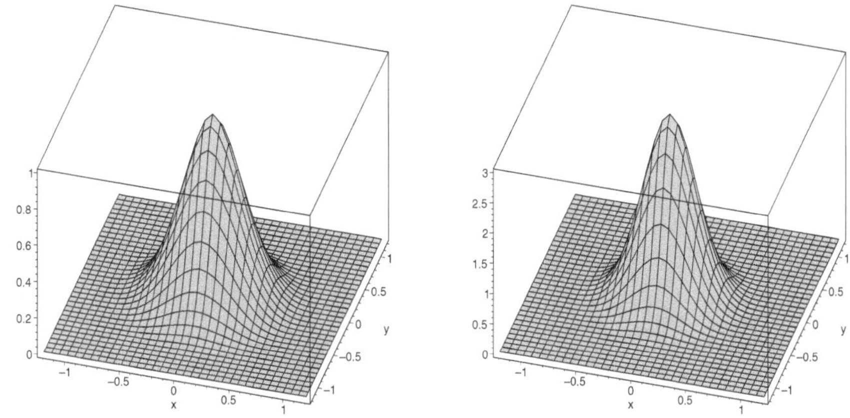

Fig. 3.1. The Wendland functions $\Psi(x) = \psi_{3,1}(\|x\|)$ (left) and $\Psi(x) = \psi_{4,2}(\|x\|)$ (right) for $x \in \mathbb{R}^2$.

$\mathcal{F}$ and its dual $\mathcal{F}^*$, cf. Section 3.2. The space $\mathcal{F}$ contains the approximated and approximating function Q, q, respectively; the corresponding dual space $\mathcal{F}^*$ includes the linear operator of the orbital derivative evaluated at a point. $\mathcal{F}^*$ is a subset of the space of distributions $\mathcal{D}'(\mathbb{R}^n)$. The native space can be characterized via Fourier transformation; the Fourier transform of the Wendland functions has been studied in [62]. In the case of the Wendland functions, the native space turns out to be the Sobolev space $\mathcal{F} = H^{\frac{n+1}{2}+k}(\mathbb{R}^n)$, where H denotes the Sobolev space, n the space dimension and $k \in \mathbb{N}$ corresponds to the respective radial basis function of the Wendland function family. In order to obtain $Q \in \mathcal{F}$, Q has to be of a certain smoothness, in our case $Q \in H^{\frac{n+1}{2}+k}(\mathbb{R}^n)$. Since Q will later be one of the Lyapunov functions T or V, f also has to have this smoothness.

The Wendland functions are positive definite functions with compact support, cf. Figure 3.1. On their support they are polynomials. They are defined by the following recursion with respect to the parameter k.

Definition 3.9 (Wendland functions). *Let $l \in \mathbb{N}$, $k \in \mathbb{N}_0$. We define by recursion*

$$\psi_{l,0}(r) = (1-r)_+^l \tag{3.17}$$

$$\textit{and } \psi_{l,k+1}(r) = \int_r^1 t\psi_{l,k}(t)\,dt \tag{3.18}$$

for $r \in \mathbb{R}_0^+$. Here we set $x_+ = x$ for $x \geq 0$ and $x_+ = 0$ for $x < 0$.

For the functions $\psi_{3,1}(cr)$, $\psi_{4,2}(cr)$ and $\psi_{5,3}(cr)$, where $c > 0$ and $\psi_{l,k}$ denotes the Wendland function, we have calculated the corresponding functions ψ_1 and ψ_2, cf. Table 3.1. Note that these are the Wendland functions of Definition 3.9 up to a constant.

	$\psi_{3,1}(cr)$	$\psi_{4,2}(cr)$
$\psi(r)$	$(1-cr)_+^4[4cr+1]$	$(1-cr)_+^6[35(cr)^2+18cr+3]$
$\psi_1(r)$	$-20c^2(1-cr)_+^3$	$-56c^2(1-cr)_+^5[1+5cr]$
$\frac{d}{dr}\psi_1(r)$	$60c^3(1-cr)_+^2$	$1680c^4r(1-cr)_+^4$
$\psi_2(r)$	$60c^3\frac{1}{r}(1-cr)_+^2$	$1680c^4(1-cr)_+^4$

	$\psi_{5,3}(cr)$
$\psi(r)$	$(1-cr)_+^8[32(cr)^3+25(cr)^2+8cr+1]$
$\psi_1(r)$	$-22c^2(1-cr)_+^7[16(cr)^2+7cr+1]$
$\frac{d}{dr}\psi_1(r)$	$528c^4r(1-cr)_+^6[6cr+1]$
$\psi_2(r)$	$528c^4(1-cr)_+^6[6cr+1]$

Table 3.1. The functions ψ_1 and ψ_2, cf. (3.7) and (3.8), for the Wendland functions $\psi_{3,1}(cr)$, $\psi_{4,2}(cr)$ and $\psi_{5,3}(cr)$. Note, that these are the Wendland functions of Definition 3.9 up to a constant. The parameters fulfill $l = k+2$. Hence, these are the appropriate Wendland functions for the dimensions $n = 2$ and $n = 3$, since here $l = \lfloor \frac{n}{2} \rfloor + k + 1 = k + 2$.

We summarize some properties of the Wendland functions in the following Proposition 3.10.

Proposition 3.10 (Properties of Wendland functions) *The Wendland function $\psi_{l,k}$ with $l \in \mathbb{N}$ and $k \in \mathbb{N}_0$ satisfies:*

1. $\operatorname{supp}(\psi_{l,k}) \subset [0,1]$. *We have $\psi_{l,k}(r) = 0$ for all $r \geq 1$, i.e. in particular $\psi_{l,k}(1) = 0$.*
2. *$\psi_{l,k}(r)$ is a polynomial of degree $l+2k$ for $r \in [0,1]$.*
3. *The function $\Psi(x) := \psi_{l,k}(\|x\|)$ is C^{2k} in 0.*
4. *$\psi_{l,k}$ is C^{k+l-1} in 1.*

PROOF: (cf. [61])

1. $k = 0$: this follows directly from the definition (3.17); $k \geq 1$: this follows from (3.18) by induction.

2. For $k = 0$ this is clear. By induction we have: if $\psi_{l,k}$ is a polynomial of degree $l+2k$, then $t\psi_{l,k}(t)$ is a polynomial of degree $l+2k+1$ and integration augments the degree to $l+2k+2$.

3. Because of 2., 3. is equivalent to: the coefficients of the first k odd powers are zero. We prove this by induction with respect to k: For $\psi_{l,0}$ there is nothing to show. Now let $k \geq 1$ and assume that $\psi_{l,k}(r) = \sum_{j=0}^{l+2k} c_j r^j$ with $c_{2i-1} = 0$ for $i = 1, \ldots, k$. Then $\psi_{l,k+1}(r) = \sum_{j=0}^{l+2k} \frac{c_j}{j+2} t^{j+2}\Big|_r^1 = C - \sum_{j=0}^{l+2k} \frac{c_j}{j+2} r^{j+2}$. The coefficient of r is zero and the coefficients of $r^3, r^5, \ldots, r^{2k+1}$ are zero by assumption.

4. $\psi_{l,0}$ is C^{l-1} in 1. Since $\frac{d}{dr}\psi_{l,k+1}(r) = -r\psi_{l,k}(r)$, the induction hypothesis implies that $\psi_{l,k}$ is C^{k+l-1} at 1. □

We fix the parameter l depending on the space dimension n and the parameter k. Then we have the following properties for the function $\Psi(x) = \psi_{l,k}(c\|x\|)$ with scaling parameter $c > 0$.

Proposition 3.11 *Let $k \in \mathbb{N}$ and $l := \left\lfloor \frac{n}{2} \right\rfloor + k + 1$. Let $\Psi(x) := \psi_{l,k}(c\|x\|)$ with $c > 0$. Then*

1. *$\psi_{l,k}(cr)$ is a polynomial of degree $\left\lfloor \frac{n}{2} \right\rfloor + 3k + 1$ for $r \in \left[0, \frac{1}{c}\right]$.*
2. *$\Psi \in C_0^{2k}(\mathbb{R}^n, \mathbb{R})$, where C_0^{2k} denotes the C^{2k} functions with compact support.*
3. *The Fourier transform $\hat{\Psi}(\omega)$ is an analytic function. We have $\hat{\Psi}(\omega) \in \mathbb{R}$ for all $\omega \in \mathbb{R}^n$ and*

$$C_1 \left(1 + \|\omega\|^2\right)^{-\frac{n+1}{2}-k} \leq \hat{\Psi}(\omega) \leq C_2 \left(1 + \|\omega\|^2\right)^{-\frac{n+1}{2}-k} \tag{3.19}$$

   with positive constants C_1, C_2. In particular, $\hat{\Psi}(\omega) > 0$ holds for all $\omega \in \mathbb{R}^n$. For the definition of the Fourier transform, cf. Appendix A.2, Definition A.11.
4. *For $\psi(r) := \psi_{l,k}(cr)$ we have the following asymptotics for $r \to 0$, for the definition of the functions ψ_1 and ψ_2 cf. (3.7) and (3.8):*
   - *$\frac{d}{dr}\psi(r) = O(r)$,*
   - *$\psi_1(r) = O(1)$ and ψ_1 can be extended continuously to 0*
   - *$\frac{d}{dr}\psi_1(r) = O(1)$ if $k = 1$ and $\frac{d}{dr}\psi_1(r) = O(r)$ if $k \geq 2$,*
   - *$\psi_2(r) = O\left(\frac{1}{r}\right)$ if $k = 1$ and $\psi_2(r) = O(1)$ if $k \geq 2$.*

Proof: 1. and 2. follow from Proposition 3.10 since $k+l-1 = \left\lfloor \frac{n}{2} \right\rfloor + 2k \geq 2k$.

We prove 3.: $\hat{\Psi}$ is analytic since Ψ has compact support, cf. Proposition A.18. First we consider the case $c = 1$ and denote the function $\Phi(x) = \psi_{l,k}(\|x\|)$. Radial functions also have a radial Fourier transform. It can be expressed by the following formula where $J_m(x)$ denotes the Bessel functions of first kind of order m: $\hat{\Phi}(\omega) = \|\omega\|^{-\frac{n-2}{2}} \int_0^\infty \psi_{l,k}(t) t^{\frac{n}{2}} J_{\frac{n-1}{2}}(\|\omega\| t)\, dt$, cf. [62]. In particular, $\hat{\Phi}$ is a real-valued function.

Moreover, cf. [61] or [62], we have $\hat{\Phi}(\omega) > 0$ for all $\omega \in \mathbb{R}^n$ and

$$C_1' \|\omega\|^{-n-2k-1} \leq \hat{\Phi}(\omega) \leq C_2' \|\omega\|^{-n-2k-1} \tag{3.20}$$

with positive constants C_1', C_2'; the first inequality only holds for $\|\omega\| \geq r_0$ with some $r_0 > 0$, the second for $\omega \neq 0$.

Now consider the Fourier transform $\hat{\Psi}(\omega)$ of $\Psi(x) = \Phi(cx) = \psi_{l,k}(\|cx\|)$ with $c > 0$. We have with the transformation $y = cx$

$$\begin{aligned}\hat{\Psi}(x) &= \int_{\mathbb{R}^n} \Phi(cx)e^{-i\langle x,\omega\rangle}\,dx \\ &= \frac{1}{c^n}\int_{\mathbb{R}^n} \Phi(y)e^{-\frac{i}{c}\langle y,\omega\rangle}\,dy \\ &= \frac{1}{c^n}\hat{\Phi}\left(\frac{\omega}{c}\right).\end{aligned}$$

The inequality (3.19) follows by (3.20).

For 4. note that by the recursion formula (3.18) we have

$$\frac{d}{dr}\psi_{l,k+1}(cr) = -c^2 r\psi_{l,k}(cr) \tag{3.21}$$

$$\psi_{1_{l,k+1}}(cr) = -c^2\psi_{l,k}(cr), \text{ cf. (3.7)} \tag{3.22}$$

for all $k \in \mathbb{N}_0$ and all $0 < r < \frac{1}{c}$. By (3.21) we have $\frac{d}{dr}\psi_{l,k}(cr) = O(r)$ for $r \to 0$ if $k \geq 1$. (3.22) shows $\psi_{1_{l,k}}(r) = O(1)$ for $r \to 0$ if $k \geq 1$, and $\psi_{1_{l,k}}(r)$ can be extended continuously to 0 by $\psi_{1_{l,k}}(0) = -c^2\psi_{l,k}(0)$.

Setting $k = 0$ in (3.22) we obtain with $\frac{d}{dr}\psi_{l,0}(cr) = \frac{d}{dr}(1-cr)^l = -cl(1-cr)^{l-1} = O(1)$ for $r \to 0$ since $l \geq 1$:

$$\frac{d}{dr}\psi_{1_{l,1}}(cr) = -c^2\frac{d}{dr}\psi_{l,0}(cr) = O(1) \text{ for } r \to 0.$$

Moreover, combining (3.22) with (3.21), we have

$$\frac{d}{dr}\psi_{1_{l,k+2}}(cr) = -c^2\frac{d}{dr}\psi_{l,k+1}(cr) = c^4 r\psi_{l,k}(cr).$$

For $k \geq 2$ we have thus

$$\frac{d}{dr}\psi_{1_{l,k}}(cr) = O(r) \text{ for } r \to 0.$$

The statements for ψ_2 follow from the definition (3.8). This shows the proposition. □

3.2 Native Space

Throughout this book we define the radial basis function Ψ by a Wendland function with parameter $k \geq 1$.

Assumption Let $k \in \mathbb{N}$, $l := \left\lfloor \frac{n}{2} \right\rfloor + k + 1$ and $c > 0$. Define the radial basis function by

$$\Psi(x) = \psi_{l,k}(c\|x\|),$$

where $\psi_{l,k}$ denotes a Wendland function, cf. Definition 3.9.

3.2.1 Characterization of the Native Space

The goal of this section is (i) to show that the interpolation matrix A of each interpolation problem is positive definite and (ii) to provide an estimate of the error of q to Q, namely $|Dq(x) - DQ(x)|$ where D is the corresponding linear operator. Note that $|Dq(x) - DQ(x)| = 0$ holds for all points $x \in X_N$ of the grid by construction. For all other points the error will depend on the density of the grid, i.e. the denser the grid is, the smaller is the error. This is done for the operator $D = D^0$ (identity), $D =$ orbital derivative and $D = D_m$ (orbital derivative and multiplication) as well as for the mixed interpolation.

In order to achieve these two goals, we define the Hilbert spaces $\mathcal{F}$ and $\mathcal{F}^*$ which are dual spaces to each other. $\mathcal{F}$ is a function space, including the function Q which will be approximated, as well as the approximating function q. The dual $\mathcal{F}^*$, on the other hand, is a space of operators, among them $(\delta_x \circ D)$, where D denotes the operator which is approximated, e.g. in our case the orbital derivative. These spaces will enable us to split the interpolation error into a part in $\mathcal{F}$ and a part in $\mathcal{F}^*$. Note that for the Wendland functions the spaces $\mathcal{F}$ and $\mathcal{F}^*$ are Sobolev spaces, cf. [61] or Lemma 3.13. For the relation of native spaces to reproducing kernel Hilbert spaces, cf. [63].

In Appendix A we summarize the results about Distributions and Fourier transformation that we need in this section. In the following definition, $\mathcal{S}'(\mathbb{R}^n)$ denotes the dual of the Schwartz space $\mathcal{S}(\mathbb{R}^n)$ of rapidly decreasing functions, cf. Appendix A.2, Definition A.17. For distributions $\lambda \in \mathcal{S}'(\mathbb{R}^n)$ we define as usual $\langle \check{\lambda}, \varphi \rangle := \langle \lambda, \check{\varphi} \rangle$ and $\langle \hat{\lambda}, \varphi \rangle := \langle \lambda, \hat{\varphi} \rangle$ with $\varphi \in C_0^\infty(\mathbb{R}^n)$, where $\check{\varphi}(x) = \varphi(-x)$ and $\hat{\varphi}(\omega) = \int_{\mathbb{R}^n} \varphi(x) e^{-ix^T\omega}\, dx$ denotes the Fourier transform.

Since Ψ is defined by a Wendland function by assumption, $\Psi \in \mathcal{E}'(\mathbb{R}^n)$, where $\mathcal{E}'(\mathbb{R}^n)$ denotes the distributions with compact support. Then the Fourier transform $\hat{\Psi}(\omega) =: \varphi(\omega)$ is an analytic function. Moreover, we have shown that $\varphi(\omega) = \hat{\Psi}(\omega) > 0$ holds for all $\omega \in \mathbb{R}^n$, cf. Proposition 3.11, 3.

Definition 3.12 (Native Space). *Let $\Psi \in \mathcal{E}'(\mathbb{R}^n)$ be defined by a Wendland function and denote its Fourier transform by $\varphi(\omega) := \hat{\Psi}(\omega)$. We define the Hilbert space*

$$\mathcal{F}^* := \left\{ \lambda \in \mathcal{S}'(\mathbb{R}^n) \mid \hat{\check{\lambda}}(\omega)\, (\varphi(\omega))^{\frac{1}{2}} \in L_2(\mathbb{R}^n) \right\},$$

with the scalar product

$$\langle \lambda, \mu \rangle_{\mathcal{F}^*} := (2\pi)^{-n} \int_{\mathbb{R}^n} \hat{\check{\lambda}}(\omega) \overline{\hat{\check{\mu}}(\omega)} \varphi(\omega)\, d\omega.$$

The native space *$\mathcal{F}$ of approximating functions is identified with the dual $\mathcal{F}^{**}$ of $\mathcal{F}^*$. The norm is given by $\|f\|_{\mathcal{F}} := \sup_{\lambda \in \mathcal{F}^*, \lambda \neq 0} \frac{|\lambda(f)|}{\|\lambda\|_{\mathcal{F}^*}}$.*

In the case of the Wendland functions, $\mathcal{F}$ and $\mathcal{F}^*$ are Sobolev spaces.

Lemma 3.13. *(cf. [61]) Since $\Psi(x) = \psi_{l,k}(c\|x\|)$ with the Wendland function, $k \in \mathbb{N}$, $l := \lfloor \frac{n}{2} \rfloor + k + 1$ and $c > 0$ we have, using the properties of the Fourier transform in Proposition 3.11,*

$$\mathcal{F}^* = H^{-\frac{n+1}{2}-k}(\mathbb{R}^n),$$

where $H^{-\frac{n+1}{2}-k}(\mathbb{R}^n)$ denotes the Sobolev space, cf. Definition A.19. The norm in Definition 3.12 and the norm of the Sobolev space $H^{-\frac{n+1}{2}-k}(\mathbb{R}^n)$ are equivalent to each other. Moreover,

$$C_0^\sigma(\mathbb{R}^n, \mathbb{R}^n) \subset \mathcal{F} = H^{\frac{n+1}{2}+k}(\mathbb{R}^n), \tag{3.23}$$

with $\mathbb{N} \ni \sigma \geq \frac{n+1}{2} + k$.

PROOF: Note that $\lambda \in H^{-\frac{n+1}{2}-k}(\mathbb{R}^n) \Leftrightarrow \check{\lambda} \in H^{-\frac{n+1}{2}-k}(\mathbb{R}^n)$. We use the estimate for the Fourier transform of the Wendland function (3.19). We have

$$\int_{\mathbb{R}^n} |\hat{\check{\lambda}}(\omega)|^2 \varphi(\omega)\, d\omega \leq C_2 \int_{\mathbb{R}^n} |\hat{\check{\lambda}}(\omega)|^2 (1 + \|\omega\|^2)^{-\frac{n+1}{2}-k}\, d\omega < \infty,$$

for $\check{\lambda} \in H^{-\frac{n+1}{2}-k}(\mathbb{R}^n)$. Thus, $\mathcal{F}^* \supset H^{-\frac{n+1}{2}-k}(\mathbb{R}^n)$.

On the other hand we have by (3.19)

$$C_1 \int_{\mathbb{R}^n} |\hat{\check{\lambda}}(\omega)|^2 (1 + \|\omega\|^2)^{-\frac{n+1}{2}-k}\, d\omega \leq \int_{\mathbb{R}^n} |\hat{\check{\lambda}}(\omega)|^2 \varphi(\omega)\, d\omega < \infty,$$

for $\check{\lambda} \in \mathcal{F}^*$. Thus, we obtain $\mathcal{F}^* \subset H^{-\frac{n+1}{2}-k}(\mathbb{R}^n)$. This shows that the norms of the respective Sobolev space and $\|\cdot\|_{\mathcal{F}}$, $\|\cdot\|_{\mathcal{F}^*}$ are equivalent to each other. Similar inequalities show (3.23). □

We have the following alternative formulas for the scalar product of $\mathcal{F}^*$, cf. [17].

Proposition 3.14 *For $\lambda, \mu \in \mathcal{F}^*$ and $\Psi \in \mathcal{E}'(\mathbb{R}^n)$ we have*

$$\begin{aligned} \langle \lambda, \mu \rangle_{\mathcal{F}^*} &= \lambda^x \overline{\mu}^y \Psi(x - y) && (3.24)\\ &= \lambda(\overline{\mu} * \Psi). && (3.25) \end{aligned}$$

PROOF: We denote the L_2 scalar product by $\langle \cdot, \cdot \rangle$. We have, cf. Appendix A.2, in particular Proposition A.18, 3. and 4., and Proposition A.10

$$\begin{aligned} \langle \lambda, \mu \rangle_{\mathcal{F}^*} &= (2\pi)^{-n} \langle \hat{\check{\lambda}}, \overline{\hat{\check{\mu}}} \hat{\Psi} \rangle \\ &= (2\pi)^{-n} \langle \hat{\check{\lambda}}, \hat{\overline{\mu}} \hat{\Psi} \rangle \\ &= (2\pi)^{-n} \check{\lambda} \widehat{\hat{\overline{\mu}} \hat{\Psi}} \\ &= \check{\lambda}(\check{\overline{\mu}} * \check{\Psi}) \\ &= \lambda(\overline{\mu} * \Psi), \text{ i.e. (3.25)} \\ &= \lambda(\overline{\mu}^y \Psi(\cdot - y)) \\ &= \lambda^x \overline{\mu}^y \Psi(x - y), \end{aligned}$$

which shows (3.24). □

Also for the norm of $\mathcal{F}$ we have a characterization via the Fourier transform, cf. (3.26). In order to prove Proposition 3.17 we recall the definition of the Riesz representative for general Hilbert spaces.

Theorem 3.15 (Riesz Theorem). *Let $\mathcal{F}$ be a Hilbert space. For every $f \in \mathcal{F}$ there is one and only one $\lambda = \lambda_f \in \mathcal{F}^*$, such that*

$$\mu(f) = \langle \mu, \lambda_f \rangle_{\mathcal{F}^*}$$

holds for all $\mu \in \mathcal{F}^*$. Moreover, $\|\lambda_f\|_{\mathcal{F}^*} = \|f\|_{\mathcal{F}}$ and $\langle f, g \rangle_{\mathcal{F}} = \langle \lambda_g, \lambda_f \rangle_{\mathcal{F}^*}$ hold.

The Riesz representative of $\overline{\mu} * \Psi$ is particularly easy to characterize.

Lemma 3.16. *Let $\mu \in \mathcal{F}^*$. Then $\mu * \Psi \in \mathcal{F}$ and $\|\mu\|_{\mathcal{F}^*} = \|\mu * \Psi\|_{\mathcal{F}}$.*
Moreover, $f := \overline{\mu} * \Psi \in \mathcal{F}$ and $\lambda_f = \mu$.

PROOF: First, we show the second part. We have for all $\lambda \in \mathcal{F}^*$

$$\langle \lambda, \lambda_f \rangle_{\mathcal{F}^*} = \lambda(f) = \lambda(\overline{\mu} * \Psi) = \langle \lambda, \mu \rangle_{\mathcal{F}^*}$$

by (3.25). This shows $\lambda_f = \mu$.

For all $\lambda \in \mathcal{F}^*$ we have $|\lambda(f)| = |\langle \lambda, \mu \rangle_{\mathcal{F}^*}| \leq \|\lambda\|_{\mathcal{F}^*} \cdot \|\mu\|_{\mathcal{F}^*}$. Thus, $f \in \mathcal{F}$ and $\|f\|_{\mathcal{F}} \leq \|\mu\|_{\mathcal{F}^*} = \|\overline{\mu}\|_{\mathcal{F}^*}$. Equality follows by choosing $\lambda = \mu$, hence $\|\overline{\mu} * \Psi\|_{\mathcal{F}} = \|\overline{\mu}\|_{\mathcal{F}^*}$.

Now we show the first part: For $\mu \in \mathcal{F}^*$, we have $\overline{\mu} \in \mathcal{F}^*$ and thus $\|\mu * \Psi\|_{\mathcal{F}} = \|\mu\|_{\mathcal{F}^*}$ with the above argumentation. □

Proposition 3.17

1. *For $f \in \mathcal{F}$ we have $\hat{f} = \overline{\hat{\lambda}_f}\varphi$, where λ_f denotes the Riesz representative, cf. Theorem 3.15.*
2. *For $f, g \in \mathcal{F}$ we have*

$$\langle f, g \rangle_{\mathcal{F}} = (2\pi)^{-n} \int_{\mathbb{R}^n} \hat{f}(\omega)\overline{\hat{g}(\omega)} \frac{1}{\varphi(\omega)} \, d\omega. \tag{3.26}$$

PROOF: 1. By the Riesz Theorem 3.15 we have for all $\lambda \in \mathcal{F}^*$

$$\begin{aligned}
\langle \lambda, \check{f} \rangle &= \check{\lambda}(f) \\
&= \langle \check{\lambda}, \lambda_f \rangle_{\mathcal{F}^*} \\
&= (2\pi)^{-n} \int_{\mathbb{R}^n} \hat{\lambda}(\omega)\overline{\hat{\lambda}_f(\omega)\varphi(\omega)} \, d\omega \\
&= (2\pi)^{-n} \langle \lambda, \widehat{\overline{\hat{\lambda}_f}\varphi} \rangle \\
\Rightarrow \check{f} &= (2\pi)^{-n} \widehat{\overline{\hat{\lambda}_f}\varphi}.
\end{aligned}$$

Hence, since $\hat{\hat{f}} = \check{\hat{f}}$, cf. Proposition A.18, 1. we have

$$\check{\hat{f}} = (\overline{\hat{\hat{\lambda}}_f \varphi})\check{} \qquad \text{and} \qquad \hat{f} = \overline{\hat{\hat{\lambda}}_f \varphi}.$$

2. Because of $\varphi(\omega) \in \mathbb{R}$ we have with Theorem 3.15

$$\begin{aligned}\langle f, g\rangle_{\mathcal{F}} &= \langle \lambda_g, \lambda_f\rangle_{\mathcal{F}^*} \\ &= (2\pi)^{-n} \int_{\mathbb{R}^n} \hat{\lambda}_g(\omega)\overline{\hat{\hat{\lambda}}_f(\omega)\varphi(\omega)}\, d\omega \\ &= (2\pi)^{-n} \int_{\mathbb{R}^n} \overline{\hat{g}(\omega)}\hat{f}(\omega)\frac{1}{\varphi(\omega)}\, d\omega \text{ by 1.}\end{aligned}$$

This shows the proposition. □

3.2.2 Positive Definiteness of the Interpolation Matrices

In this section we prove the positive definiteness of the interpolation matrices. More precisely, there are four different interpolation problems with corresponding interpolation matrices, cf. Sections 3.1.1, 3.1.2 and 3.1.3. Note that the function Ψ throughout the rest of the book is given by $\Psi(x) = \psi_{l,k}(c\|x\|)$ where $\psi_{l,k}$ is the Wendland function with parameters $k \in \mathbb{N}$ and $l = \lfloor \frac{n}{2} \rfloor + k + 1$; $c > 0$ is a scaling parameter. First we define the spaces of approximating functions $\mathcal{F}_{X_M^0}$, $\mathcal{F}_{X_N}$, $\mathcal{F}_{X_N, X_M^0}$, respectively, and show that they are subsets of $\mathcal{F}$. Afterwards, we show the positive definiteness.

Approximation via Function Values

The space $\mathcal{F}^*_{X_M^0}$ in the following definition includes the evaluation of the function at a grid point. The space $\mathcal{F}_{X_M^0}$ includes the approximating functions.

Definition 3.18. *Let $X_M^0 = \{\xi_1, \xi_2, \ldots, \xi_M\}$ be a set of pairwise distinct points. We define*

$$\begin{aligned}\mathcal{F}^*_{X_M^0} &:= \left\{\lambda \in \mathcal{S}'(\mathbb{R}^n) \mid \lambda = \sum_{j=1}^M \gamma_j \delta_{\xi_j}, \gamma_j \in \mathbb{R}\right\} \\ \mathcal{F}_{X_M^0} &:= \{\lambda * \Psi \mid \lambda \in \mathcal{F}^*_{X_M^0}\},\end{aligned}$$

where $\Psi(x) = \psi_{l,k}(c\|x\|)$ with the Wendland function $\psi_{l,k}$, where $k \in \mathbb{N}$ and $l = \lfloor \frac{n}{2} \rfloor + k + 1$.

Lemma 3.19. *We have $\mathcal{F}^*_{X_M^0} \subset \mathcal{F}^* \cap \mathcal{E}'(\mathbb{R}^n)$ and $\mathcal{F}_{X_M^0} \subset \mathcal{F}$.*

PROOF: For $\lambda \in \mathcal{F}^*_{X^0_M}$ we have $\text{supp}(\lambda) = \cup_{j=1}^M \{\xi_j\} =: K$ and

$$\hat{\lambda}(\omega) = \sum_{j=1}^M \gamma_j \delta^x_{\xi_j} e^{ix^T\omega} = \sum_{j=1}^M \gamma_j e^{i\xi_j^T\omega}.$$

With Proposition 3.11 we have

$$\begin{aligned}\int_{\mathbb{R}^n} |\hat{\lambda}(\omega)|^2 \varphi(\omega)\, d\omega &\leq C_2 \left(\sum_{j=1}^M |\gamma_j|\right)^2 \int_{\mathbb{R}^n} \left(1 + \|\omega\|^2\right)^{-\frac{n+1}{2}-k} d\omega \\ &\leq C \int_{\mathbb{R}^n} \left(1 + \|\omega\|^2\right)^{-\frac{n+1}{2}} d\omega < \infty.\end{aligned}$$

This shows $\mathcal{F}^*_{X_N} \subset \mathcal{F}^*$. The second inclusion follows by Lemma 3.16. □

Proposition 3.20 *Let $\Psi(x) = \psi_{l,k}(c\|x\|)$ with the Wendland function $\psi_{l,k}$, where $k \in \mathbb{N}$ and $l = \lfloor \frac{n}{2} \rfloor + k + 1$. Let $X^0_M = \{\xi_1, \xi_2, \ldots, \xi_M\}$ be a set of pairwise distinct points. Then the interpolation matrix A^0 defined in Definition 3.1 is positive definite.*

PROOF: For $\lambda = \sum_{i=1}^M \gamma_i \delta_{\xi_i} \in \mathcal{F}^* \cap \mathcal{E}'(\mathbb{R}^n)$ we have with Proposition 3.14 $\gamma^T A^0 \gamma = \lambda^x \overline{\lambda}^y \Psi(x-y) = \|\lambda\|^2_{\mathcal{F}^*} = (2\pi)^{-n} \int_{\mathbb{R}^n} |\hat{\lambda}(\omega)|^2 \varphi(\omega)\, d\omega$. Since $\varphi(\omega) = \hat{\Psi}(\omega) > 0$ holds for all $\omega \in \mathbb{R}^n$, cf. Proposition 3.11, the matrix A^0 is positive semidefinite.

Now we show that $\gamma^T A^0 \gamma = 0$ implies $\gamma = 0$. We assume that $\gamma^T A^0 \gamma = (2\pi)^{-n} \int_{\mathbb{R}^n} |\hat{\lambda}(\omega)|^2 \varphi(\omega)\, d\omega = 0$. Then the analytic function satisfies $\hat{\lambda}(\omega) = 0$ for all $\omega \in \mathbb{R}^n$. By Fourier transformation in $\mathcal{S}'(\mathbb{R}^n)$ we have $\mathcal{S}'(\mathbb{R}^n) \ni \lambda = 0$, i.e.

$$\lambda(h) = \sum_{i=1}^M \gamma_i h(\xi_i) \; = \; 0 \tag{3.27}$$

for all test functions $h \in \mathcal{S}(\mathbb{R}^n)$. Fix a $j \in \{1, \ldots, M\}$. Since the points ξ_i are distinct, there is a neighborhood $B_\delta(\xi_j)$ such that $\xi_i \notin B_\delta(\xi_j)$ holds for all $i \neq j$. Define the function $h(x) = 1$ for $x \in B_{\frac{\delta}{2}}(\xi_j)$ and $h(x) = 0$ for $x \notin B_\delta(\xi_j)$, and extend it smoothly such that $h \in \mathcal{S}(\mathbb{R}^n)$. Then (3.27) yields

$$0 = \lambda(h) = \gamma_j.$$

This argumentation holds for all $j = 1, \ldots, M$ and thus $\gamma = 0$. Hence, A^0 is positive definite. □

Approximation via Orbital Derivatives

The space $\mathcal{F}^*_{X_N}$ in the following definition includes the evaluation of the orbital derivative at a grid point. The space $\mathcal{F}_{X_N}$ includes the approximating functions.

Definition 3.21. *Let $X_N = \{x_1, x_2, \ldots, x_N\}$ be a set of pairwise distinct points, which are no equilibrium points. We define*

$$\mathcal{F}^*_{X_N} := \left\{ \lambda \in \mathcal{S}'(\mathbb{R}^n) \mid \lambda = \sum_{j=1}^{N} \beta_j (\delta_{x_j} \circ D), \beta_j \in \mathbb{R} \right\}$$

$$\mathcal{F}_{X_N} := \{\lambda * \Psi \mid \lambda \in \mathcal{F}^*_{X_N}\}$$

where $Dq(x) = q'(x)$ denotes the orbital derivative and $\Psi(x) = \psi_{l,k}(c\|x\|)$ with the Wendland function $\psi_{l,k}$, where $k \in \mathbb{N}$ and $l = \lfloor \frac{n}{2} \rfloor + k + 1$.

Lemma 3.22. *We have $\mathcal{F}^*_{X_N} \subset \mathcal{F}^* \cap \mathcal{E}'(\mathbb{R}^n)$ and $\mathcal{F}_{X_N} \subset \mathcal{F}$.*
Moreover, $\delta_x \in \mathcal{F}^* \cap \mathcal{E}'(\mathbb{R}^n)$ for all $x \in \mathbb{R}^n$.

PROOF: For $\lambda \in \mathcal{F}^*_{X_N}$ we have $\operatorname{supp}(\lambda) = \cup_{j=1}^N \{x_j\} =: K$ and

$$\hat{\hat{\lambda}}(\omega) = \sum_{j=1}^{N} \beta_j (\delta_{x_j} \circ D)^x e^{ix^T\omega} = i \sum_{j=1}^{N} \beta_j \langle \omega, f(x_j) \rangle e^{ix_j^T\omega}.$$

With Proposition 3.11 we have

$$\begin{aligned} &\int_{\mathbb{R}^n} |\hat{\hat{\lambda}}(\omega)|^2 \varphi(\omega)\, d\omega \\ &\leq C_2 \left(\sum_{j=1}^{N} |\beta_j| \max_{x \in K} \|f(x)\| \right)^2 \int_{\mathbb{R}^n} \|\omega\|^2 \left(1 + \|\omega\|^2\right)^{-\frac{n+1}{2}-k} d\omega \\ &\leq C \int_{\mathbb{R}^n} \left(1 + \|\omega\|^2\right)^{-\frac{n+1}{2}} d\omega < \infty \end{aligned}$$

since $k \geq 1$. This shows $\mathcal{F}^*_{X_N} \subset \mathcal{F}^*$. The second inclusion follows by Lemma 3.16.

For $x \in \mathbb{R}^n$ we have $\operatorname{supp}(\delta_x) = \{x\}$ and

$$\hat{\hat{\delta}}_x(\omega) = e^{ix^T\omega}.$$

With Proposition 3.11 we have

$$\int_{\mathbb{R}^n} |\hat{\hat{\delta}}_x(\omega)|^2 \varphi(\omega)\, d\omega \leq C_2 \int_{\mathbb{R}^n} \left(1 + \|\omega\|^2\right)^{-\frac{n+1}{2}-k} d\omega < \infty.$$

□

Proposition 3.23 *Let $\Psi(x) = \psi_{l,k}(c\|x\|)$ with the Wendland function $\psi_{l,k}$, where $k \in \mathbb{N}$ and $l = \lfloor \frac{n}{2} \rfloor + k + 1$. Let $X_N = \{x_1, x_2, \ldots, x_N\}$ be a set of pairwise distinct points, which are no equilibrium points. Then the matrix A of Proposition 3.5 is positive definite.*

PROOF: For $\lambda = \sum_{i=1}^N \beta_i(\delta_{x_i} \circ D) \in \mathcal{F}^* \cap \mathcal{E}'(\mathbb{R}^n)$ we have with Proposition 3.14 $\beta^T A \beta = \lambda^x \overline{\lambda}^y \Psi(x-y) = \|\lambda\|^2_{\mathcal{F}^*} = (2\pi)^{-n} \int_{\mathbb{R}^n} |\hat{\lambda}(\omega)|^2 \varphi(\omega)\, d\omega$. Since $\varphi(\omega) = \hat{\Psi}(\omega) > 0$ holds for all $\omega \in \mathbb{R}^n$, cf. Proposition 3.11, the matrix A is positive semidefinite.

Now we show that $\beta^T A \beta = 0$ implies $\beta = 0$. We assume that $\beta^T A \beta = (2\pi)^{-n} \int_{\mathbb{R}^n} |\hat{\lambda}(\omega)|^2 \varphi(\omega)\, d\omega = 0$. Then the analytic function satisfies $\hat{\lambda}(\omega) = 0$ for all $\omega \in \mathbb{R}^n$. By Fourier transformation in $\mathcal{S}'(\mathbb{R}^n)$ we have $\mathcal{S}'(\mathbb{R}^n) \ni \lambda = 0$, i.e.

$$\lambda(h) = \sum_{i=1}^N \beta_i \langle \nabla h(x_i), f(x_i) \rangle \;=\; 0 \tag{3.28}$$

for all test functions $h \in \mathcal{S}(\mathbb{R}^n)$. Fix a $j \in \{1, \ldots, N\}$. Since the points x_i are distinct, there is a neighborhood $B_\delta(x_j)$ such that $x_i \notin B_\delta(x_j)$ holds for all $i \neq j$. Define the function $h(x) = \langle x - x_j, f(x_j) \rangle$ for $x \in B_{\frac{\delta}{2}}(x_j)$ and $h(x) = 0$ for $x \notin B_\delta(x_j)$, and extend it smoothly such that $h \in \mathcal{S}(\mathbb{R}^n)$. Then (3.28) yields

$$0 = \lambda(h) = \beta_j \|f(x_j)\|^2.$$

Since x_j is no equilibrium, $\beta_j = 0$. This argumentation holds for all $j = 1, \ldots, N$ and thus $\beta = 0$. Hence, A is positive definite. □

Approximation via Orbital Derivatives and Multiplication

In the following definition the space $\mathcal{F}_{X_N}$ includes the approximating functions.

Definition 3.24. *Let $X_N = \{x_1, x_2, \ldots, x_N\}$ be a set of pairwise distinct points, which are no equilibrium points. Let $m\colon \mathbb{R}^n \setminus \{x_0\} \to \mathbb{R}$ be continuous in $\mathbb{R}^n \setminus \{x_0\}$ and bounded in $B_\epsilon(x_0) \setminus \{x_0\}$ for each $\epsilon > 0$. Let $D = D_m$ be given by $D_m Q(x) := Q'(x) + m(x)Q(x)$.*

We define

$$\mathcal{F}^*_{X_N} := \left\{ \lambda \in \mathcal{S}'(\mathbb{R}^n) \mid \lambda = \sum_{j=1}^N \beta_j(\delta_{x_j} \circ D_m), \beta_j \in \mathbb{R} \right\}$$

$$\mathcal{F}_{X_N} := \{\lambda * \Psi \mid \lambda \in \mathcal{F}^*_{X_N}\},$$

where $\Psi(x) = \psi_{l,k}(c\|x\|)$ with the Wendland function $\psi_{l,k}$, where $k \in \mathbb{N}$ and $l = \lfloor \frac{n}{2} \rfloor + k + 1$.

Lemma 3.25. *We have* $\mathcal{F}^*_{X_N} \subset \mathcal{F}^* \cap \mathcal{E}'(\mathbb{R}^n)$ *and* $\mathcal{F}_{X_N} \subset \mathcal{F}$.
Moreover, $\delta_x \in \mathcal{F}^* \cap \mathcal{E}'(\mathbb{R}^n)$ *for all* $x \in \mathbb{R}^n$.

PROOF: For $\lambda \in \mathcal{F}^*_{X_N}$ we have $\operatorname{supp}(\lambda) = \cup_{j=1}^N \{x_j\} =: K$; note that $x_0 \notin K$. We have

$$\hat{\lambda}(\omega) = \sum_{j=1}^N \beta_j (\delta_{x_j} \circ D_m)^x e^{ix^T\omega} = \sum_{j=1}^N \beta_j \left(i\langle \omega, f(x_j)\rangle + m(x_j)\right) e^{ix_j^T\omega}.$$

Since m is continuous in $\mathbb{R}^n \setminus \{x_0\}$, there is a constant $M > 0$ such that $|m(x)| \leq M$ holds for all $x \in K$. With Proposition 3.11 we have

$$\begin{aligned}
&\int_{\mathbb{R}^n} |\hat{\lambda}(\omega)|^2 \varphi(\omega)\, d\omega \\
&\leq 2C_2 \left(\sum_{j=1}^N |\beta_j| \max_{x\in K} \|f(x)\| \right)^2 \int_{\mathbb{R}^n} \|\omega\|^2 \left(1 + \|\omega\|^2\right)^{-\frac{n+1}{2}-k} d\omega \\
&\quad +2C_2 \left(\sum_{j=1}^N |\beta_j| \max_{x\in K} |m(x)| \right)^2 \int_{\mathbb{R}^n} \left(1 + \|\omega\|^2\right)^{-\frac{n+1}{2}-k} d\omega \\
&\leq C \int_{\mathbb{R}^n} \left(1 + \|\omega\|^2\right)^{-\frac{n+1}{2}} d\omega \\
&< \infty
\end{aligned}$$

since $k \geq 1$. This shows $\mathcal{F}^*_{X_N} \subset \mathcal{F}^*$. The second inclusion follows by Lemma 3.16. For $\delta_x \in \mathcal{F}^* \cap \mathcal{E}'(\mathbb{R}^n)$, cf. Lemma 3.22. □

Proposition 3.26 *Let* $\Psi(x) = \psi_{l,k}(c\|x\|)$ *with the Wendland function* $\psi_{l,k}$, *where* $k \in \mathbb{N}$ *and* $l = \lfloor \frac{n}{2} \rfloor + k + 1$. *Let* $X_N = \{x_1, x_2, \ldots, x_N\}$ *be a set of pairwise distinct points, which are no equilibrium points. Then the matrix* A *defined in Proposition 3.6 is positive definite.*

PROOF: For $\lambda = \sum_{i=1}^N \beta_i (\delta_{x_i} \circ D_m) \in \mathcal{F}^* \cap \mathcal{E}'(\mathbb{R}^n)$ we have with Proposition 3.14 $\beta^T A \beta = \lambda^x \overline{\lambda}^y \Psi(x-y) = \|\lambda\|^2_{\mathcal{F}^*} = (2\pi)^{-n} \int_{\mathbb{R}^n} |\hat{\lambda}(\omega)|^2 \varphi(\omega)\, d\omega$. Since $\varphi(\omega) = \hat{\Psi}(\omega) > 0$ holds for all $\omega \in \mathbb{R}^n$, the matrix A is positive semidefinite.

Now we show that $\beta^T A \beta = 0$ implies $\beta = 0$. We assume that $\beta^T A \beta = (2\pi)^{-n} \int_{\mathbb{R}^n} |\hat{\lambda}(\omega)|^2 \varphi(\omega)\, d\omega = 0$. Then the analytic function satisfies $\hat{\lambda}(\omega) = 0$ for all $\omega \in \mathbb{R}^n$. By Fourier transformation in $\mathcal{S}'(\mathbb{R}^n)$ we have $\mathcal{S}'(\mathbb{R}^n) \ni \lambda = 0$, i.e.

$$\lambda(h) = \sum_{i=1}^N \beta_i \left[\langle \nabla h(x_i), f(x_i)\rangle + m(x_i) h(x_i) \right] = 0 \tag{3.29}$$

for all test functions $h \in \mathcal{S}(\mathbb{R}^n)$. Fix a $j \in \{1, \ldots, N\}$. Since the points x_i are distinct, there is a neighborhood $B_\delta(x_j)$ such that $x_i \notin B_\delta(x_j)$ holds for all $i \neq j$. Define the function $h(x) = \langle x - x_j, f(x_j)\rangle$ for $x \in B_{\frac{\delta}{2}}(x_j)$ and $h(x) = 0$ for $x \notin B_\delta(x_j)$, and extend it smoothly such that $h \in \mathcal{S}(\mathbb{R}^n)$. Then (3.29) yields

$$0 = \lambda(h) = \beta_j \|f(x_j)\|^2.$$

Since x_j is no equilibrium, $\beta_j = 0$. This argumentation holds for all $j = 1, \ldots, N$ and thus $\beta = 0$. Hence, A is positive definite. □

Mixed Approximation

In the case of mixed approximation we define the space $\mathcal{F}_{X_N, X_M^0}$, which includes the approximating functions, and the space $\mathcal{F}^*_{X_N, X_M^0}$, which includes the operators of the orbital derivative evaluated at a grid point of X_N and the operators of the evaluation at a grid point of X_M^0.

Definition 3.27. *For the grids $X_N = \{x_1, x_2, \ldots, x_N\}$ with operator $Dq(x) = q'(x)$ (orbital derivative) and $X_M^0 = \{\xi_1, \ldots, \xi_M\}$ with operator $D^0 = \mathrm{id}$, such that all x_i are no equilibria, we define*

$$\mathcal{F}^*_{X_N, X_M^0} := \left\{ \lambda \in \mathcal{S}'(\mathbb{R}^n) \mid \lambda = \sum_{j=1}^N \beta_j (\delta_{x_j} \circ D) + \sum_{i=1}^M \gamma_i (\delta_{\xi_i} \circ D^0) \right\}$$

$$\mathcal{F}_{X_N, X_M^0} := \{\lambda * \Psi \mid \lambda \in \mathcal{F}^*_{X_N, X_M^0}\},$$

where $\beta_j, \gamma_i \in \mathbb{R}$ and $\Psi(x) = \psi_{l,k}(c\|x\|)$ with the Wendland function $\psi_{l,k}$, where $k \in \mathbb{N}$ and $l = \lfloor \frac{n}{2} \rfloor + k + 1$.

Lemma 3.28. *We have $\mathcal{F}^*_{X_N, X_M^0} \subset \mathcal{F}^* \cap \mathcal{E}'(\mathbb{R}^n)$ and $\mathcal{F}_{X_N, X_M^0} \subset \mathcal{F}$.*

Proof: For $\lambda \in \mathcal{F}^*_{X_N, X_M^0}$ we have $\mathrm{supp}(\lambda) = \bigcup_{j=1}^N \{x_j\} \cup \bigcup_{j=1}^M \{\xi_j\} =: K$ and

$$\begin{aligned} \hat{\lambda}(\omega) &= \sum_{j=1}^N \beta_j (\delta_{x_j} \circ D)^x e^{ix^T\omega} + \sum_{j=1}^M \gamma_j e^{i\xi_j^T\omega} \\ &= i \sum_{j=1}^N \beta_j \langle \omega, f(x_j)\rangle e^{ix_j^T\omega} + \sum_{j=1}^M \gamma_j e^{i\xi_j^T\omega}. \end{aligned}$$

With Proposition 3.11 we have

$$\begin{aligned}
&\int_{\mathbb{R}^n} |\hat{\lambda}(\omega)|^2 \varphi(\omega)\, d\omega \\
&\leq 2C_2 \left(\sum_{j=1}^{N} |\beta_j| \max_{x\in K} \|f(x)\| \right)^2 \int_{\mathbb{R}^n} \|\omega\|^2 \left(1+\|\omega\|^2\right)^{-\frac{n+1}{2}-k} d\omega \\
&\quad +2C_2 \left(\sum_{j=1}^{M} |\gamma_j| \right)^2 \int_{\mathbb{R}^n} \left(1+\|\omega\|^2\right)^{-\frac{n+1}{2}-k} d\omega \\
&\leq C \int_{\mathbb{R}^n} \left(1+\|\omega\|^2\right)^{-\frac{n+1}{2}} d\omega \\
&< \infty
\end{aligned}$$

since $k \geq 1$. This shows $\mathcal{F}^*_{X_N, X_M^0} \subset \mathcal{F}^*$. The second inclusion follows again with Lemma 3.16. □

The positive definiteness of the corresponding interpolation matrix is shown in the following proposition. Note that X_N and X_M^0 may have common points.

Proposition 3.29 *Let $\Psi(x) = \psi_{l,k}(c\|x\|)$ with the Wendland function $\psi_{l,k}$, where $k \in \mathbb{N}$ and $l = \lfloor \frac{n}{2} \rfloor + k + 1$. Let X_N and X_M^0 be grids as in Definition 3.27, i.e. X_N contains no equilibria.*

Then the interpolation matrix $\begin{pmatrix} A & C \\ C^T & A^0 \end{pmatrix}$, cf. Definition 3.7 and Proposition 3.8, is positive definite.

PROOF: For $\lambda = \sum_{i=1}^N \beta_i(\delta_{x_i} \circ D) + \sum_{i=1}^M \gamma_i(\delta_{\xi_i} \circ D^0) \in \mathcal{F}^* \cap \mathcal{E}'(\mathbb{R}^n)$ (by Lemma 3.28) we have $(\beta, \gamma) \begin{pmatrix} A & C \\ C^T & A^0 \end{pmatrix} \begin{pmatrix} \beta \\ \gamma \end{pmatrix} = \lambda^x \overline{\lambda}^y \Psi(x-y) = \|\lambda\|^2_{\mathcal{F}^*} = (2\pi)^{-n} \int_{\mathbb{R}^n} |\hat{\lambda}(\omega)|^2 \varphi(\omega)\, d\omega$. Since $\varphi(\omega) > 0$ holds for all ω, the matrix is positive semidefinite.

Now we show that $(\beta, \gamma) \begin{pmatrix} A & C \\ C^T & A^0 \end{pmatrix} \begin{pmatrix} \beta \\ \gamma \end{pmatrix} = 0$ implies $\beta = 0$ and $\gamma = 0$. If $(\beta, \gamma) \begin{pmatrix} A & C \\ C^T & A^0 \end{pmatrix} \begin{pmatrix} \beta \\ \gamma \end{pmatrix} = (2\pi)^{-n} \int_{\mathbb{R}^n} |\hat{\lambda}(\omega)|^2 \varphi(\omega)\, d\omega = 0$, then the analytic function satisfies $\hat{\lambda}(\omega) = 0$ for all $\omega \in \mathbb{R}^n$. By Fourier transformation in $\mathcal{S}'(\mathbb{R}^n)$ we have $\mathcal{S}'(\mathbb{R}^n) \ni \lambda = 0$, i.e.

$$\lambda(h) = \sum_{i=1}^{N} \beta_i \langle \nabla h(x_i), f(x_i) \rangle + \sum_{i=1}^{M} \gamma_i h(\xi_i) \;=\; 0 \tag{3.30}$$

for all test functions $h \in \mathcal{S}(\mathbb{R}^n)$. Fix a $j \in \{1, \ldots, N\}$. Either there is a point $\xi_{j^*} = x_j$ with $j^* \in \{1, \ldots, M\}$; then there is a neighborhood $B_\delta(x_j)$ such that $x_i \notin B_\delta(x_j)$ holds for all $i \neq j$ and $\xi_i \notin B_\delta(x_j)$ holds for all $i \neq j^*$. Otherwise we can choose $B_\delta(x_j)$ such that $x_i \notin B_\delta(x_j)$ holds for all $i \neq j$ and $\xi_i \notin B_\delta(x_j)$

holds for all i. In both cases define the function $h(x) = \langle x - x_j, f(x_j)\rangle$ for $x \in B_{\frac{\delta}{2}}(x_j)$ and $h(x) = 0$ for $x \notin B_\delta(x_j)$, and extend it smoothly such that $h \in \mathcal{S}(\mathbb{R}^n)$. Then (3.30) yields in both cases

$$0 = \lambda(h) = \beta_j \|f(x_j)\|^2.$$

Since x_j is no equilibrium, $\beta_j = 0$. This argumentation holds for all $j = 1, \ldots, N$ and thus $\beta = 0$.

Now fix a $j \in \{1, \ldots, M\}$. Either there is a point $x_{j^*} = \xi_j$ with $j^* \in \{1, \ldots, N\}$; then there is a neighborhood $B_\delta(\xi_j)$ such that $\xi_i \notin B_\delta(\xi_j)$ holds for all $i \neq j$ and $x_i \notin B_\delta(\xi_j)$ holds for all $i \neq j^*$. Otherwise we can choose $B_\delta(\xi_j)$ such that $\xi_i \notin B_\delta(\xi_j)$ holds for all $i \neq j$ and $x_i \notin B_\delta(\xi_j)$ holds for all i. In both cases define the function $h(x) = 1$ for $x \in B_{\frac{\delta}{2}}(\xi_j)$ and $h(x) = 0$ for $x \notin B_\delta(\xi_j)$, and extend it smoothly such that $h \in \mathcal{S}(\mathbb{R}^n)$. Then (3.30) yields in both cases

$$0 = \lambda(h) = \gamma_j.$$

This argumentation holds for all $j = 1, \ldots, M$ and thus $\gamma = 0$.

Hence, the matrix $\begin{pmatrix} A & C \\ C^T & A^0 \end{pmatrix}$ is positive definite. □

3.2.3 Error Estimates

While the interpolation of function values and the corresponding error estimates have been studied in detail, the interpolation via the values of a linear operator is relatively new. The values of such linear operators are also called Hermite-Birkhoff data. The mixed problem occurs when solving a Cauchy problem in partial differential equations. This has been done in Franke & Schaback [17] and [18]. Their error estimates used the fact that the linear operator is translation invariant and, hence, they studied a partial differential equation with constant coefficients. Wendland [63] considered the case of non-constant coefficients without zeros. Our linear operator D, however, is the orbital derivative, namely $Dq(x) = q'(x) = \langle \nabla q(x), f(x)\rangle$, and D is not translation invariant. The coefficients $f_i(x)$ are neither constant nor nonzero. Hence, we have to use different techniques for our error estimates, which are Taylor expansions in this book. In [27], one obtains better approximation orders using a general convergence result [51], but one has to make further assumptions on the set K. We prefer the direct estimates using the Taylor expansion in this book.

In this section we estimate the error between the approximated function Q and the approximant q. Depending on the ansatz we can give a bound on the corresponding operator evaluated at any point in a compact set, i.e. if we approximate Q via function values, we obtain an estimate on the values $|Q(x) - q(x)|$, if we approximate Q via the orbital derivative, we obtain a bound on the orbital derivative $|Q'(x) - q'(x)|$. For the mixed approximation both errors within the respective sets can be estimated.

We fix a compact set. The denser the grid is within this compact set, the smaller is the error. We define the fill distance.

Definition 3.30 (Fill distance). *Let $K \subset \mathbb{R}^n$ be a compact set and let $X_N := \{x_1, \ldots, x_N\} \subset K$ be a grid. The positive real number*

$$h := h_{K,X_N} = \max_{y\in K} \min_{x\in X_N} \|x - y\|$$

is called the fill distance of X_N in K.

In particular, for all $y \in K$ there is a grid point $x_k \in X_N$ such that $\|y - x_k\| \leq h$.

For $\lambda = \delta_x \circ D$, where δ denotes Dirac's δ-distribution and D a linear operator, we will use the estimate $|\lambda(Q) - \lambda(q)| \leq |(\lambda - \mu)(Q - q)| \leq \|\lambda - \mu\|_{\mathcal{F}^*} \|Q - q\|_{\mathcal{F}}$, where $\mu \in \mathcal{F}^*_{X^0_M}, \mathcal{F}^*_{X_N}, \mathcal{F}^*_{X_N,X^0_M}$, respectively. Note that $\mu(Q - q) = 0$ since $(DQ)(x_j) = (Dq)(x_j)$ holds for all grid points x_j. For the second term we prove $\|Q - q\|_{\mathcal{F}} \leq \|Q\|_{\mathcal{F}}$ and for the first term we use a Taylor expansion. We distinguish again between the four different cases.

Approximation via Function Values

Proposition 3.31 *Let $\Psi(x) = \psi_{l,k}(c\|x\|)$ with the Wendland function $\psi_{l,k}$, where $k \in \mathbb{N}$ and $l = \lfloor \frac{n}{2} \rfloor + k + 1$. Let $X^0_M = \{\xi_1, \xi_2, \ldots, \xi_M\}$ be a set of pairwise distinct points. Let $Q \in \mathcal{F}$ and q be the reconstruction of Q with respect to the grid X^0_M in the sense of Definition 3.1, i.e. $q(x) = \sum_{j=1}^M \gamma_j \Psi(x - \xi_j) \in \mathcal{F}_{X^0_M}$ and γ is the solution of $A^0\gamma = \alpha^0$ with $\alpha^0_j = Q(\xi_j)$. Then*

$$\|Q - q\|_{\mathcal{F}} \leq \|Q\|_{\mathcal{F}}.$$

Proof: Note that Q and q are real-valued functions. We have $q = \lambda * \Psi$ with $\lambda = \sum_{j=1}^M \gamma_j \delta_{\xi_j} = \overline{\lambda} \in \mathcal{F}^* \cap \mathcal{E}'(\mathbb{R}^n)$. Thus,

$$\begin{aligned}
\|Q - q\|^2_{\mathcal{F}} &= \|Q\|^2_{\mathcal{F}} + \|q\|^2_{\mathcal{F}} - 2\langle Q, q\rangle_{\mathcal{F}} \\
&= \|Q\|^2_{\mathcal{F}} + \|\lambda * \Psi\|^2_{\mathcal{F}} - 2\langle \lambda_q, \lambda_Q\rangle_{\mathcal{F}^*}, \text{ cf. Theorem 3.15} \\
&= \|Q\|^2_{\mathcal{F}} + \|\lambda\|^2_{\mathcal{F}^*} - 2\langle \lambda, \lambda_Q\rangle_{\mathcal{F}^*}, \text{ cf. Lemma 3.16} \\
&= \|Q\|^2_{\mathcal{F}} + \lambda^x \overline{\lambda}^y \Psi(x - y) - 2\lambda(Q), \\
&\qquad \text{cf. Proposition 3.14 and Theorem 3.15} \\
&= \|Q\|^2_{\mathcal{F}} + \gamma^T A^0 \gamma - 2\gamma^T \alpha^0 \\
&= \|Q\|^2_{\mathcal{F}} - \gamma^T A^0 \gamma \\
&\leq \|Q\|^2_{\mathcal{F}},
\end{aligned}$$

since A^0 is positive (semi)definite, cf. Proposition 3.20. □

Now we prove an error estimate.

Theorem 3.32. *Consider the function $\Psi(x) = \psi_{l,k}(c\|x\|)$ with $c > 0$, where $\psi_{l,k}$ denotes the Wendland function with $k \in \mathbb{N}$ and $l := \lfloor \frac{n}{2} \rfloor + k + 1$.*

Let K be a compact set, and $Q \in \mathcal{F}$. For all $H_0 > 0$ there is a constant $C_0^* = C_0^*(K)$ such that: Let $X_M^0 := \{\xi_1, \ldots, \xi_M\} \subset K$ be a grid with fill distance $0 < h_0 < H_0$ in K and let $q \in C^{2k}(\mathbb{R}^n, \mathbb{R})$ be the reconstruction of Q with respect to the grid X_M^0. Then

$$|Q(x) - q(x)| \leq C_0^* h_0 \text{ holds for all } x \in K. \tag{3.31}$$

PROOF: Let $x^* \in K$ and set $\lambda = \delta_{x^*} \in \mathcal{F}^*$. This follows in a similar way as in Lemma 3.19. Let $\mu \in \mathcal{F}^*_{X_M^0}$; then $\mu(Q - q) = 0$. In order to prove (3.31) – note, that the left hand sides of (3.32) and (3.31) are equal for $x = x^*$ – we have thus

$$\begin{aligned} |\lambda(Q) - \lambda(q)| = |(\lambda - \mu)(Q - q)| &\leq \|\lambda - \mu\|_{\mathcal{F}^*} \cdot \|Q - q\|_{\mathcal{F}} \\ &\leq \|\lambda - \mu\|_{\mathcal{F}^*} \cdot \|Q\|_{\mathcal{F}} \end{aligned} \tag{3.32}$$

by Proposition 3.31. Now choose $\mu = \delta_{\tilde{x}}$ with $\tilde{x} = \xi_i$ for an $i \in \{1, \ldots, M\}$ such that $\|x^* - \tilde{x}\| \leq h_0$. Then we have, denoting $\psi(r) = \psi_{l,k}(cr)$,

$$\begin{aligned} \|\lambda - \mu\|_{\mathcal{F}^*}^2 &= (\lambda - \mu)^x (\lambda - \mu)^y \Psi(x - y) \\ &= (\delta_{x^*} - \delta_{\tilde{x}})^x (\delta_{x^*} - \delta_{\tilde{x}})^y \Psi(x - y) \\ &= 2[\psi(0) - \psi(\|x^* - \tilde{x}\|)]. \end{aligned}$$

Denoting $r := \|x^* - \tilde{x}\| \leq h_0$ and using the Taylor expansion there is an $\tilde{r} \in [0, r]$ such that

$$\begin{aligned} \|\lambda - \mu\|_{\mathcal{F}^*}^2 &= -2 \frac{d}{dr} \psi(\tilde{r}) r \\ &= O(r^2) \text{ for } r \to 0 \end{aligned}$$

by the property $\frac{d}{dr}\psi(r) = O(r)$ of the Wendland functions, cf. 4. of Proposition 3.11. Hence, $\|\lambda - \mu\|_{\mathcal{F}^*} = O(r)$ and we obtain (3.31) by (3.32). This proves the proposition. □

Remark 3.33 *Using Fourier analysis, one can show a higher convergence rate. Applying [68], Theorems 4 and 5, we have $s_\infty = 2k+1$ for the Wendland functions, and thus $|Q(x) - q(x)| \leq C_0^* (h_0)^{k+\frac{1}{2}}$ for all $x \in K$. One can transfer this result to operators which are translation invariant, cf. [17]. However, since the orbital derivative is not translation invariant, this is not possible in our case. Moreover, for the mixed approximation we have an ansatz for q involving a non-translation invariant part. Hence, for the following results, we can estimate the error in a similar way as in Theorem 3.32. Note that higher convergence orders for the orbital derivative have been obtained in [27] by a different method.*

Approximation via Orbital Derivatives

Proposition 3.34 *Let $\Psi(x) = \psi_{l,k}(c\|x\|)$ with the Wendland function $\psi_{l,k}$, where $k \in \mathbb{N}$ and $l = \lfloor \frac{n}{2} \rfloor + k + 1$. Let $X_N = \{x_1, x_2, \ldots, x_N\}$ be a set of pairwise distinct points, which are no equilibrium points. Let $Q \in \mathcal{F}$ and q be the reconstruction of Q with respect to the grid X_N in the sense of Proposition 3.5, i.e. $q(x) = \sum_{j=1}^N \beta_j (\delta_{x_j} \circ D)^y \Psi(x-y) \in \mathcal{F}_{X_N}$ and β is the solution of $A\beta = \alpha$ with $\alpha_j = (\delta_{x_j} \circ D)^x Q(x)$, where $Dq(x) = q'(x)$ denotes the orbital derivative. Then*

$$\|Q - q\|_{\mathcal{F}} \leq \|Q\|_{\mathcal{F}}.$$

PROOF: Note that Q and q are real-valued functions. We have $q = \lambda * \Psi$ with $\lambda = \sum_{j=1}^N \beta_j (\delta_{x_j} \circ D) = \overline{\lambda} \in \mathcal{F}^* \cap \mathcal{E}'(\mathbb{R}^n)$. Thus,

$$\begin{aligned}
\|Q - q\|_{\mathcal{F}}^2 &= \|Q\|_{\mathcal{F}}^2 + \|q\|_{\mathcal{F}}^2 - 2\langle Q, q\rangle_{\mathcal{F}} \\
&= \|Q\|_{\mathcal{F}}^2 + \|\lambda * \Psi\|_{\mathcal{F}}^2 - 2\langle \lambda_q, \lambda_Q\rangle_{\mathcal{F}^*}, \text{ cf. Theorem 3.15} \\
&= \|Q\|_{\mathcal{F}}^2 + \|\lambda\|_{\mathcal{F}^*}^2 - 2\langle \lambda, \lambda_Q\rangle_{\mathcal{F}^*}, \text{ cf. Lemma 3.16} \\
&= \|Q\|_{\mathcal{F}}^2 + \lambda^x \overline{\lambda}^y \Psi(x-y) - 2\lambda(Q), \\
&\qquad\qquad \text{cf. Proposition 3.14 and Theorem 3.15} \\
&= \|Q\|_{\mathcal{F}}^2 + \beta^T A \beta - 2\beta^T \alpha \\
&= \|Q\|_{\mathcal{F}}^2 - \beta^T A \beta \\
&\leq \|Q\|_{\mathcal{F}}^2,
\end{aligned}$$

since A is positive (semi)definite, cf. Proposition 3.23. □

Now we give an error estimate in Theorem 3.35. Note that the convergence rate κ depends on the parameter k of the Wendland functions.

Theorem 3.35. *Consider the function $\Psi(x) = \psi_{l,k}(c\|x\|)$ with $c > 0$, where $\psi_{l,k}$ denotes the Wendland function with $k \in \mathbb{N}$ and $l := \lfloor \frac{n}{2} \rfloor + k + 1$. Let $f \in C^1(\mathbb{R}^n, \mathbb{R}^n)$.*

Let K be a compact set, and $Q \in \mathcal{F}$. For all $H > 0$ there is a constant $C^* = C^*(K)$ such that: Let $X_N := \{x_1, \ldots, x_N\} \subset K$ be a grid with pairwise distinct points which are no equilibria and with fill distance $0 < h < H$ in K, and let $q \in C^{2k-1}(\mathbb{R}^n, \mathbb{R})$ be the reconstruction of Q with respect to the grid X_N and operator $Dq(x) = q'(x)$ (orbital derivative).

Then

$$|Q'(x) - q'(x)| \leq C^* h^\kappa \text{ holds for all } x \in K, \tag{3.33}$$

where $\kappa = \frac{1}{2}$ for $k = 1$ and $\kappa = 1$ for $k \geq 2$.

PROOF: Let $x^* \in K$ and set $\lambda = \delta_{x^*} \circ D \in \mathcal{F}^*$; this follows in a similar way as in Lemma 3.22. Let $\mu \in \mathcal{F}_{X_N}^*$. We have $\mu(Q - q) = 0$. Hence,

$$\begin{aligned}|\lambda(Q)-\lambda(q)| = |(\lambda-\mu)(Q-q)| &\leq \|\lambda-\mu\|_{\mathcal{F}^*}\cdot\|Q-q\|_{\mathcal{F}}\\ &\leq \|\lambda-\mu\|_{\mathcal{F}^*}\cdot\|Q\|_{\mathcal{F}}\end{aligned}\tag{3.34}$$

by Proposition 3.34. We choose $\mu = \delta_{\tilde{x}} \circ D$ with $\tilde{x} = x_i$ for an $i \in \{1, \ldots, N\}$ such that $\|x^* - \tilde{x}\| \leq h$. Then we have with a similar calculation as in Proposition 3.5

$$\begin{aligned}\|\lambda-\mu\|^2_{\mathcal{F}^*} &= (\lambda-\mu)^x(\lambda-\mu)^y\Psi(x-y)\\ &= (\delta_{x^*}\circ D - \delta_{\tilde{x}}\circ D)^x(\delta_{x^*}\circ D - \delta_{\tilde{x}}\circ D)^y\Psi(x-y)\\ &= -\psi_1(0)\|f(x^*)\|^2 - \psi_1(0)\|f(\tilde{x})\|^2\\ &\quad -2\psi_2(\|x^*-\tilde{x}\|)\langle x^*-\tilde{x}, f(x^*)\rangle\langle \tilde{x}-x^*, f(\tilde{x})\rangle\\ &\quad +2\psi_1(\|x^*-\tilde{x}\|)\|\langle f(x^*), f(\tilde{x})\rangle\\ &\leq -\psi_1(0)\|f(x^*)-f(\tilde{x})\|^2\\ &\quad +2\left[\psi_1(\|x^*-\tilde{x}\|)-\psi_1(0)\right]\langle f(x^*), f(\tilde{x})\rangle\\ &\quad +2|\psi_2(\|x^*-\tilde{x}\|)|\cdot\|x^*-\tilde{x}\|^2\cdot\|f(x^*)\|\cdot\|f(\tilde{x})\|.\end{aligned}$$

There are constants c_0, c_1 such that $\|f(x)\| \leq c_0$ and $\|Df(x)\| \leq c_1$ hold for all $x \in \overline{\text{conv}(K)}$. Denoting $r := \|x^* - \tilde{x}\| \leq h$ and using Taylor's Theorem there are $\tilde{r} \in [0, r]$ and $\xi = \theta x^* + (1-\theta)\tilde{x}$ where $\theta \in [0, 1]$ such that

$$\|\lambda-\mu\|^2_{\mathcal{F}^*} \leq |\psi_1(0)|\underbrace{\|Df(\xi)\|^2}_{\leq c_1^2} r^2 + 2c_0^2\left|\frac{d}{dr}\psi_1(\tilde{r})\right| r + 2c_0^2\,|\psi_2(r)|\,r^2 \tag{3.35}$$

holds. Thus, by the asymptotics of Proposition 3.11, 4. the right hand side of (3.35) is of order $O(r)$ for $k = 1$ and of order $O(r^2)$ for $k \geq 2$. Hence, we have $\|\lambda - \mu\|_{\mathcal{F}^*} \leq \tilde{C}h^\kappa$ for $h < H$, where $\kappa = \frac{1}{2}$ for $k = 1$ and $\kappa = 1$ for $k \geq 2$. Thus, (3.33) follows by (3.34) with $C^* := \tilde{C}\|Q\|_{\mathcal{F}}$. □

Remark 3.36 *The error estimate in (3.33) can be improved concerning the order of convergence κ depending on k using recent results in [51] and [63], cf. [27]. There, the convergence in (3.33) is $h^{k-\frac{1}{2}}$ instead of h^κ.*

Approximation via Orbital Derivatives and Multiplication

Proposition 3.37 *Let $\Psi(x) = \psi_{l,k}(c\|x\|)$ with the Wendland function $\psi_{l,k}$, where $k \in \mathbb{N}$ and $l = \lfloor \frac{n}{2} \rfloor + k + 1$. Let $X_N = \{x_1, x_2, \ldots, x_N\}$ be a set of pairwise distinct points, which are no equilibrium points. Let $Q \in \mathcal{F}$ and q be the reconstruction of Q with respect to the grid X_N in the sense of Proposition 3.6, i.e. $q(x) = \sum_{j=1}^N \beta_j(\delta_{x_j} \circ D_m)^y\Psi(x-y) \in \mathcal{F}_{X_N}$ and β is the solution of $A\beta = \alpha$ with $\alpha_j = (\delta_{x_j} \circ D_m)^x Q(x)$. Then*

$$\|Q - q\|_{\mathcal{F}} \leq \|Q\|_{\mathcal{F}}.$$

The proof is similar to the one of Proposition 3.34.

With the function $\mathfrak{n}(x)$ defined in Definition 2.56 we set $m(x) := \frac{\mathfrak{n}'(x)}{\mathfrak{n}(x)}$. In the following proposition we show several properties of m and related functions that we need for the proof of Theorem 3.39.

Proposition 3.38 *Let $\sigma \geq P \geq 2$ and let $\mathfrak{n}(x)$ be as in Definition 2.56; let $n \geq 2$. We define the functions $m \in C^\sigma(\mathbb{R}^n \setminus \{x_0\}, \mathbb{R})$, $m^* \in C^\infty(\mathbb{R}^n \setminus \{x_0\}, \mathbb{R})$ and $e \in C^\sigma(\mathbb{R}^n \setminus \{x_0\}, \mathbb{R})$ by*

$$m(x) = \frac{\mathfrak{n}'(x)}{\mathfrak{n}(x)} \tag{3.36}$$

$$m^*(x) = -\frac{\|x - x_0\|^2}{(x - x_0)^T B(x - x_0)} \tag{3.37}$$

$$e(x) = m(x) - m^*(x). \tag{3.38}$$

B denotes the matrix such that $\mathfrak{v}(x) = (x - x_0)^T B(x - x_0)$, where $\mathfrak{v}(x)$ are the terms of $\mathfrak{n}(x)$ of order two, cf. Remark 2.34 and Section 2.3.4. These functions have the following properties:

- *There is an $\epsilon > 0$ and a $C'_M > 0$ such that $|m(x)| \leq C'_M$ holds for all $x \in B_\epsilon(x_0) \setminus \{x_0\}$, i.e. m is bounded for $x \to x_0$.*
- *We have*

$$e(x) = o(1) \qquad \text{for } x \to x_0 \tag{3.39}$$

  and we set $e(x_0) := 0$ such that $e\colon \mathbb{R}^n \to \mathbb{R}$ is continuous.
- *$m^*(x) = m^*\left(x_0 + \frac{x - x_0}{\|x - x_0\|}\right)$ holds for all $x \in \mathbb{R}^n \setminus \{x_0\}$. Therefore, one can define $m^*(\varphi)$ as a C^∞-function with respect to the angle $\varphi \in [0, 2\pi) \times [0, \pi]^{n-2} =: S$ where $\varphi(x)\colon \mathbb{R}^n \setminus \{x_0\} \to S$ is the canonical mapping from $x \in \mathbb{R}^n \setminus \{x_0\}$ to the corresponding angle $\varphi \in S$. More precisely, x and $\varphi \in S$ satisfy*

$$\begin{aligned} x = x_0 + r(&\sin\varphi_1 \cdot \ldots \cdot \sin\varphi_{n-1}, \cos\varphi_1 \sin\varphi_2 \cdot \ldots \cdot \sin\varphi_{n-1}, \\ &\cos\varphi_2 \sin\varphi_3 \cdot \ldots \cdot \sin\varphi_{n-1}, \ldots, \cos\varphi_{n-1}) \end{aligned}$$

  with some $r > 0$. We have $C := \sup_{\varphi \in S} |\nabla_\varphi m^*(\varphi)| < \infty$.
  Let $X_N = \{x_1, \ldots, x_N\} \subset \mathbb{R}^n \setminus \{x_0\}$ be a grid such that the corresponding grid of the angles $\Xi_N = \{\varphi(x_1), \ldots, \varphi(x_N)\} \subset S$ is a grid in $\overline{S}$ with fill distance h_φ. Then for all $x \in \mathbb{R}^n \setminus \{x_0\}$ there is a grid point $x_i \in X_N$ such that

$$|m^*(x) - m^*(x_i)| \leq C h_\varphi \text{ holds.} \tag{3.40}$$

Proof: Note that $\mathfrak{n}'(x) = -\|x - x_0\|^2 + \varphi(x)$ where $\varphi(x) = o(\|x - x_0\|^P)$ by (2.38) and (2.28). φ is a C^σ-function since $\mathfrak{n}'(x)$ is a C^σ-function. $\mathfrak{n}(x)$, on the other hand, is given by

$$\mathfrak{n}(x) = \underbrace{(x-x_0)^T B(x-x_0)}_{=\mathfrak{v}(x)} + \underbrace{\sum_{3\le|\alpha|\le P} c_\alpha (x-x_0)^\alpha + M\|x-x_0\|^{2H}}_{=:r(x)},$$

where $r(x) = O(\|x-x_0\|^3)$ for $x \to x_0$, and $r \in C^\infty(\mathbb{R}^n, \mathbb{R}^n)$.

Now we show that m is bounded near x_0.

$$\begin{aligned} m(x) &= \frac{\mathfrak{n}'(x)}{\mathfrak{n}(x)} \\ &= \frac{-\|x-x_0\|^2}{(x-x_0)^T B(x-x_0) + r(x)} + \underbrace{\frac{\varphi(x)}{(x-x_0)^T B(x-x_0) + r(x)}}_{=o(\|x-x_0\|^{P-2})}. \end{aligned} \tag{3.41}$$

Since B is symmetric and positive definite, it has a smallest eigenvalue $\lambda > 0$. Since $r(x) = O(\|x-x_0\|^3)$, there is an ϵ such that $|r(x)| \le \frac{\lambda}{2}\|x-x_0\|^2$ holds for all $x \in B_\epsilon(x_0)$. Since $P \ge 2$, the second term of (3.41) is $o(1)$, and thus there is a constant $c^* > 0$ such that $\left|\frac{\varphi(x)}{(x-x_0)^T B(x-x_0)+r(x)}\right| \le c^*$ holds for all $x \in B_\epsilon(x_0) \setminus \{x_0\}$. Altogether, by (3.41) we have

$$|m(x)| \le \frac{\|x-x_0\|^2}{\left(\lambda - \frac{\lambda}{2}\right)\|x-x_0\|^2} + c^* = \frac{2}{\lambda} + c^* =: C'_M$$

for all $x \in B_\epsilon(x_0) \setminus \{x_0\}$.

With $m^*(x) = -\frac{\|x-x_0\|^2}{(x-x_0)^T B(x-x_0)}$ for $x \neq x_0$ we have

$$\begin{aligned} e(x) &= m(x) - m^*(x) \\ &= \frac{-\|x-x_0\|^2 + \varphi(x)}{(x-x_0)^T B(x-x_0) + r(x)} + \frac{\|x-x_0\|^2}{(x-x_0)^T B(x-x_0)} \\ &= \frac{\varphi(x)(x-x_0)^T B(x-x_0) + \|x-x_0\|^2 r(x)}{[(x-x_0)^T B(x-x_0) + r(x)](x-x_0)^T B(x-x_0)} \\ &= o(\|x-x_0\|^{P-2}) + O(\|x-x_0\|). \end{aligned}$$

Since $P \ge 2$,

$$e(x) = o(1)$$

which shows (3.39). In particular, by setting $e(x_0) = 0$, e is a continuous function in $\mathbb{R}^n$.

Using spherical coordinates with center x_0, m^* only depends on the angles $\varphi_1, \ldots, \varphi_{n-1}$ and not on the radius. Thus, we can define $m^*(\varphi)$ for $\varphi \in S$; m^* is a C^∞-function with respect to φ; this can easily be checked considering the concrete form of m^*. For $v, w \in \mathbb{R}^n \setminus \{x_0\}$ we consider the corresponding angles $\varphi(v), \varphi(w) \in S$. By Taylor's Theorem there is a $\theta \in [0,1]$ such that with $\tilde{\varphi} = \theta\varphi(v) + (1-\theta)\varphi(w)$ we have

$$\begin{aligned}|m^*(v) - m^*(w)| &= |m^*(\varphi(v)) - m^*(\varphi(w))| \\ &= |\nabla_\varphi m^*(\tilde{\varphi})(\varphi(v) - \varphi(w))| \\ &\leq C|\varphi(v) - \varphi(w)|\end{aligned}$$

since $\overline{S}$ is convex. If the grid Ξ_N has fill distance h_φ in $\overline{S}$, then for all $\varphi(v)$ there is a grid point $w = x_i$ such that $|\varphi(v) - \varphi(w)| \leq h_\varphi$. This shows (3.40). □

Now we prove Theorem 3.39 giving an error estimate for the values $D_m q(x)$. Note, that the convergence rate κ depends on the parameter k of the Wendland functions. The grid has again an upper bound for the fill distance in K and additionally a special structure is needed near x_0: here the corresponding grid Ξ_N of the angles in $S = [0, 2\pi) \times [0, \pi]^{n-2}$ has an upper bound on the fill distance in S.

Theorem 3.39. *Consider the function $\Psi(x) = \psi_{l,k}(c\|x\|)$ with $c > 0$, where $\psi_{l,k}$ denotes the Wendland function with $k \in \mathbb{N}$ and $l := \lfloor \frac{n}{2} \rfloor + k + 1$. Let $f \in C^1(\mathbb{R}^n, \mathbb{R}^n)$.*

Let K be a compact set and let $m(x) = \frac{\mathtt{n}'(x)}{\mathtt{n}(x)}$ with $\sigma \geq P \geq 2$, cf. Definition 2.56, and $Q \in \mathcal{F}$.

For all $\tilde{c} > 0$ there is a grid $X_N := \{x_1, \ldots, x_N\} \subset K \setminus \{x_0\}$ with pairwise distinct points which are no equilibria such that the reconstruction $q \in C^{2k-1}(\mathbb{R}^n, \mathbb{R})$ of Q with respect to the grid X_N and operator $D_m q(x) = q'(x) + m(x)q(x)$ satisfies

$$|D_m Q(x) - D_m q(x)| \leq \tilde{c} \quad \textit{for all } x \in K \setminus \{x_0\}. \tag{3.42}$$

Proof: Denote $\psi(r) := \psi_{l,k}(cr)$. Choose $\epsilon > 0$ such that for $e(x) = m(x) - m^*(x)$ defined in Proposition 3.38

$$|e(x)| \leq \frac{1}{8} \frac{\tilde{c}}{\|Q\|_{\mathcal{F}} \sqrt{|\psi(0)|}} \tag{3.43}$$

holds for all $x \in B_{2\epsilon}(x_0)$; note that in (3.39) we have shown that $e(x) = o(1)$ holds for $x \to x_0$.

There are constants $c_0, c_1, C_M > 0$ such that $\|f(x)\| \leq c_0$ and $\|Df(x)\| \leq c_1$ holds for all $x \in \overline{\text{conv}(K)}$, and $|m(x)| \leq C_M$ holds for all $x \in \overline{\text{conv}(K)} \setminus \{x_0\}$, cf. Proposition 3.38. Moreover, there is a constant c_M such that $|\nabla m(x)| \leq c_M$ holds for all $x \in \overline{\text{conv}(K)} \setminus B_\epsilon(x_0)$.

Choose $0 < h \leq \epsilon$ such that

$$\begin{aligned}\frac{1}{2} \frac{\tilde{c}^2}{\|Q\|_{\mathcal{F}}^2} \geq{}& |\psi_1(0)| c_1^2 h^2 + 2c_0^2 \max_{\tilde{h} \in [0,h]} \left| \frac{d}{dr} \psi_1(\tilde{h}) \right| h + 2c_0^2 \max_{\tilde{h} \in [0,h]} |\psi_2(\tilde{h})| h^2 \\ &+ 2C_M^2 \max_{\tilde{h} \in [0,h]} \left| \frac{d}{dr} \psi(\tilde{h}) \right| h + 4C_M c_0 \max_{\tilde{h} \in [0,h]} \left| \psi_1(\tilde{h}) \right| h\end{aligned} \tag{3.44}$$

holds (this is possible by the asymptotics of Proposition 3.11, 4.) and, at the same time,

$$h \leq \frac{1}{\sqrt{2|\psi(0)|}} \frac{\tilde{c}}{\|Q\|_{\mathcal{F}} c_M}. \tag{3.45}$$

Choose a grid in $K \setminus B_{2\epsilon}(x_0)$ with fill distance $\leq h$. Inside $B_{2\epsilon}(x_0)$ choose a grid in the following way: For each $x^* \in B_{2\epsilon}(x_0) \setminus \{x_0\}$ there is a grid point $\tilde{x} \in B_{2\epsilon}(x_0) \setminus \{x_0\}$ such that (i) $\|x^* - \tilde{x}\| \leq h$ and (ii) the angles have a distance $\leq h_\varphi \leq \frac{1}{4C} \frac{\tilde{c}}{\|Q\|_{\mathcal{F}} \sqrt{|\psi(0)|}}$ in S where C was defined in Proposition 3.38. Let X_N be the union of both grids.

Hence, for all points $x \in \mathbb{R}^n \setminus \{x_0\}$ there is a grid point $x_j = x_0 + \rho w_j$ with $0 < \rho < 2\epsilon$ and $w_j \in S^{n-1}$ such that

$$|m^*(x) - m^*(x_j)| \leq \frac{1}{4} \frac{\tilde{c}}{\|Q\|_{\mathcal{F}} \sqrt{|\psi(0)|}} \tag{3.46}$$

holds, cf. (3.40).

Let $x^* \in K \setminus \{x_0\}$ and set $\lambda = \delta_{x^*} \circ D_m \in \mathcal{F}^*$; this follows in a similar way as in Lemma 3.25. For $\mu \in \mathcal{F}^*_{X_N}$ we have $\mu(Q - q) = 0$ and hence

$$\begin{aligned} |\lambda(Q) - \lambda(q)| &= |(\lambda - \mu)(Q - q)| \leq \|\lambda - \mu\|_{\mathcal{F}^*} \cdot \|Q - q\|_{\mathcal{F}} \\ &\leq \|\lambda - \mu\|_{\mathcal{F}^*} \cdot \|Q\|_{\mathcal{F}} \end{aligned} \tag{3.47}$$

by Proposition 3.37. We choose $\mu = \delta_{\tilde{x}} \circ D_m$ with $\tilde{x} = x_i$ for an $i \in \{1, \ldots, N\}$ such that $\|x^* - \tilde{x}\| \leq h$. Then we have with a similar calculation as in Proposition 3.6

$$\begin{aligned} &\|\lambda - \mu\|^2_{\mathcal{F}^*} \\ &= (\lambda - \mu)^x (\lambda - \mu)^y \Psi(x - y) \\ &= (\delta_{x^*} \circ D_m - \delta_{\tilde{x}} \circ D_m)^x (\delta_{x^*} \circ D_m - \delta_{\tilde{x}} \circ D_m)^y \Psi(x - y) \\ &= -\psi_1(0)\|f(x^*)\|^2 - \psi_1(0)\|f(\tilde{x})\|^2 + \psi(0)[m(x^*)^2 + m(\tilde{x})^2] \\ &\quad -2\psi_2(\|x^* - \tilde{x}\|)\langle x^* - \tilde{x}, f(x^*)\rangle\langle \tilde{x} - x^*, f(\tilde{x})\rangle \\ &\quad + 2\psi_1(\|x^* - \tilde{x}\|)\|\langle f(x^*), f(\tilde{x})\rangle \\ &\quad - 2\psi_1(\|x^* - \tilde{x}\|)[m(x^*)\langle \tilde{x} - x^*, f(\tilde{x})\rangle + m(\tilde{x})\langle x^* - \tilde{x}, f(x^*)\rangle] \\ &\quad - 2m(x^*)m(\tilde{x})\psi(\|x^* - \tilde{x}\|) \\ &\leq -\psi_1(0)\|f(x^*) - f(\tilde{x})\|^2 \\ &\quad + 2\left[\psi_1(\|x^* - \tilde{x}\|) - \psi_1(0)\right]\langle f(x^*), f(\tilde{x})\rangle \\ &\quad + 2|\psi_2(\|x^* - \tilde{x}\|)| \cdot \|x^* - \tilde{x}\|^2 \cdot \|f(x^*)\| \cdot \|f(\tilde{x})\| \\ &\quad + \psi(0)[m(x^*) - m(\tilde{x})]^2 - 2m(x^*)m(\tilde{x})[\psi(\|x^* - \tilde{x}\|) - \psi(0)] \\ &\quad -2\psi_1(\|x^* - \tilde{x}\|)[m(x^*)\langle \tilde{x} - x^*, f(\tilde{x})\rangle + m(\tilde{x})\langle x^* - \tilde{x}, f(x^*)\rangle]. \end{aligned}$$

Denoting $r := \|x^* - \tilde{x}\| \leq h$ and using the Taylor expansion, there are $\tilde{r}_1, \tilde{r}_2 \in [0, r]$ and $\xi = \theta x^* + (1 - \theta)\tilde{x}$ where $\theta \in [0, 1]$ such that

$$\|\lambda-\mu\|_{\mathcal{F}^*}^2 \leq |\psi_1(0)| \underbrace{\|Df(\xi)\|^2}_{\leq c_1^2} r^2 + 2c_0^2 \left|\frac{d}{dr}\psi_1(\tilde{r}_1)\right| r + 2c_0^2 |\psi_2(r)| r^2$$
$$+2C_M^2 \left|\frac{d}{dr}\psi(\tilde{r}_2)\right| r + 4C_M c_0 |\psi_1(r)| r + |\psi(0)|[m(x^*) - m(\tilde{x})]^2$$
$$\leq \frac{1}{2}\frac{\tilde{c}^2}{\|Q\|_{\mathcal{F}}^2} + |\psi(0)|[m(x^*) - m(\tilde{x})]^2 \text{ by (3.44).} \tag{3.48}$$

If $x^* \in B_{2\epsilon}(x_0)$, then we can choose $\tilde{x}$ such that, additionally to $\|x^* - \tilde{x}\| \leq h$, also $\tilde{x} \in B_{2\epsilon}(x_0)$ and $\tilde{x} = x_0 + \rho w_j$ hold with $0 < \rho < 2\epsilon$ and $w_j \in S^{n-1}$. Hence, by (3.38) we have $m(x^*) - m(\tilde{x}) = m^*(x^*) - m^*(\tilde{x}) + e(x^*) - e(\tilde{x})$, and with (3.43) and (3.48)

$$\|\lambda-\mu\|_{\mathcal{F}^*}^2 \leq \frac{1}{2}\frac{\tilde{c}^2}{\|Q\|_{\mathcal{F}}^2} + |\psi(0)| \left[|m^*(x^*) - m^*(\tilde{x})| + 2\frac{1}{8}\frac{\tilde{c}}{\|Q\|_{\mathcal{F}}\sqrt{|\psi(0)|}}\right]^2$$
$$\leq \frac{1}{2}\frac{\tilde{c}^2}{\|Q\|_{\mathcal{F}}^2} + |\psi(0)| \left[\frac{1}{4}\frac{\tilde{c}}{\|Q\|_{\mathcal{F}}\sqrt{|\psi(0)|}} + \frac{1}{4}\frac{\tilde{c}}{\|Q\|_{\mathcal{F}}\sqrt{|\psi(0)|}}\right]^2 \text{ by (3.46)}$$
$$\leq \frac{\tilde{c}^2}{\|Q\|_{\mathcal{F}}^2}.$$

Thus, (3.42) follows by (3.47).

Now assume $x^* \notin B_{2\epsilon}(x_0)$. Since $h \leq \epsilon$, the line from $\tilde{x}$ to x^* has no points in common with $B_\epsilon(x_0)$. Hence, there is a $\theta \in [0,1]$ such that with $\tilde{\xi} = \theta x^* + (1-\theta)\tilde{x} \notin B_\epsilon(x_0)$ we have using (3.48)

$$\|\lambda-\mu\|_{\mathcal{F}^*}^2 \leq \frac{1}{2}\frac{\tilde{c}^2}{\|Q\|_{\mathcal{F}}^2} + |\psi(0)|[|\nabla m(\tilde{\xi})|r]^2$$
$$\leq \frac{1}{2}\frac{\tilde{c}^2}{\|Q\|_{\mathcal{F}}^2} + |\psi(0)|(c_M h)^2$$
$$\leq \frac{\tilde{c}^2}{\|Q\|_{\mathcal{F}}^2} \text{ by (3.45).}$$

Thus, (3.42) follows again by (3.47). □

Mixed Approximation

Proposition 3.40 *Let $\Psi(x) = \psi_{l,k}(c\|x\|)$ with the Wendland function $\psi_{l,k}$, where $k \in \mathbb{N}$ and $l = \lfloor \frac{n}{2} \rfloor + k + 1$. Let $X_N = \{x_1, x_2, \ldots, x_N\}$ and $X_M^0 = \{\xi_1, \ldots, \xi_M\}$ be sets of pairwise distinct points x_i, ξ_j, respectively. Let x_i be no equilibrium points. Let $Q \in \mathcal{F}$ and q be the reconstruction of Q with respect to the grids X_N, X_M^0 in the sense of Definition 3.7, i.e. $q(x) = \sum_{j=1}^N \beta_j(\delta_{x_j} \circ$*

$D)^y\Psi(x-y)+\sum_{j=1}^M \gamma_j\Psi(x-\xi_j) \in \mathcal{F}_{X_N,X_M^0}$ *and* $\begin{pmatrix}\beta\\ \gamma\end{pmatrix}$ *is the solution of* $\begin{pmatrix} A & C \\ C^T & A^0\end{pmatrix}\begin{pmatrix}\beta\\ \gamma\end{pmatrix} = \begin{pmatrix}\alpha\\ \alpha^0\end{pmatrix}$ *with* $\alpha_j = (\delta_{x_j}\circ D)^x Q(x)$ *and* $\alpha_j^0 = Q(\xi_j)$. *Here,* $Dq(x) = q'(x)$ *denotes the orbital derivative. Then*

$$\|Q-q\|_{\mathcal{F}} \le \|Q\|_{\mathcal{F}}.$$

PROOF: Note that Q and q are real-valued functions. We have $q=\lambda*\Psi$ with $\lambda = \sum_{j=1}^N \beta_j(\delta_{x_j}\circ D)+\sum_{j=1}^M \gamma_j(\delta_{\xi_j}\circ D^0) = \overline{\lambda}\in\mathcal{F}^*\cap\mathcal{E}'(\mathbb{R}^n)$, where $D^0=\text{id}$. Thus,

$$\begin{aligned}
\|Q-q\|_{\mathcal{F}}^2 &= \|Q\|_{\mathcal{F}}^2 + \|q\|_{\mathcal{F}}^2 - 2\langle Q,q\rangle_{\mathcal{F}}\\
&= \|Q\|_{\mathcal{F}}^2 + \|\lambda*\Psi\|_{\mathcal{F}}^2 - 2\langle \lambda_q,\lambda_Q\rangle_{\mathcal{F}^*}, \text{ cf. Theorem 3.15}\\
&= \|Q\|_{\mathcal{F}}^2 + \|\lambda\|_{\mathcal{F}^*}^2 - 2\langle \lambda,\lambda_Q\rangle_{\mathcal{F}^*}, \text{ cf. Lemma 3.16}\\
&= \|Q\|_{\mathcal{F}}^2 + \lambda^x\overline{\lambda}^y\Psi(x-y) - 2\lambda(Q),\\
&\qquad\qquad \text{cf. Proposition 3.14 and Theorem 3.15}\\
&= \|Q\|_{\mathcal{F}}^2 + (\beta^T,\gamma^T)\begin{pmatrix} A & C \\ C^T & A^0\end{pmatrix}\begin{pmatrix}\beta\\ \gamma\end{pmatrix} - 2(\beta^T,\gamma^T)\begin{pmatrix}\alpha\\ \alpha^0\end{pmatrix}\\
&= \|Q\|_{\mathcal{F}}^2 - (\beta^T,\gamma^T)\begin{pmatrix} A & C \\ C^T & A^0\end{pmatrix}\begin{pmatrix}\beta\\ \gamma\end{pmatrix}\\
&\le \|Q\|_{\mathcal{F}}^2
\end{aligned}$$

since $\begin{pmatrix} A & C \\ C^T & A^0\end{pmatrix}$ is positive (semi)definite, cf. Proposition 3.29. □

Theorem 3.41. *Consider the function* $\Psi(x) = \psi_{l,k}(c\|x\|)$ *with* $c>0$, *where* $\psi_{l,k}$ *denotes the Wendland function with* $k\in\mathbb{N}$ *and* $l := \lfloor \frac{n}{2}\rfloor + k + 1$. *Assume that* $f\in C^1(\mathbb{R}^n,\mathbb{R}^n)$. *Denote by* $Dq(x)=q'(x)$ *the orbital derivative.*

Let K *and* Ω *be compact sets, and let* $Q\in\mathcal{F}$. *For all* $H, H_0 > 0$ *there are constants* $C^* = C^*(K)$ *and* $C_0^* = C_0^*(\Omega)$ *such that: Let* $X_N := \{x_1,\ldots,x_N\}\subset K$ *for* $i=1,\ldots,N$ *be a grid with pairwise distinct points which are no equilibrium points and with fill distance* $0<h<H$ *in* K. *Let* $X_M^0 := \{\xi_1,\ldots,\xi_M\}\subset\Omega$ *be a grid with pairwise distinct points and fill distance* $0<h_0<H_0$ *in* Ω. *Let* $q\in C^{2k-1}(\mathbb{R}^n,\mathbb{R})$ *be the reconstruction of* Q *with respect to the grids* X_N, X_M^0, *cf. Definition 3.7. Then*

$$|Q'(x)-q'(x)| \le C^*h^\kappa \text{ holds for all } x\in K \tag{3.49}$$

$$|Q(x)-q(x)| \le C_0^*h_0 \text{ holds for all } x\in\Omega, \tag{3.50}$$

where $\kappa=\frac{1}{2}$ *for* $k=1$ *and* $\kappa=1$ *for* $k\ge 2$.

PROOF: We can follow Theorems 3.35 and 3.32. In order to prove (3.49) let $x^*\in K$ and $\tilde{x}\in X_N$ such that $r := \|x^*-\tilde{x}\|\le h$. Set $\lambda = \delta_{x^*}\circ D\in\mathcal{F}^*$, which follows in a similar way as in Lemma 3.28, and $\mu = \delta_{\tilde{x}} \circ D \in \mathcal{F}^*_{X_N, X_M^0} \subset \mathcal{F}^*$, cf. Lemma 3.28. Then we have $\mu(Q - q) = 0$ and together with Proposition 3.40

$$\begin{aligned} |\lambda(Q - q)| &= |(\lambda - \mu)(Q - q)| \leq \|\lambda - \mu\|_{\mathcal{F}^*} \cdot \|Q - q\|_{\mathcal{F}} \\ &\leq \|\lambda - \mu\|_{\mathcal{F}^*} \cdot \|Q\|_{\mathcal{F}}. \end{aligned}$$

Now we proceed as in the proof of Theorem 3.35 and obtain (3.49), cf. (3.33).

In order to prove (3.50), let $x^* \in \Omega$ and $\tilde{x} \in X_M^0$ such that $r := \|x^* - \tilde{x}\| \leq h_0$. Set $\lambda = \delta_{x^*} \in \mathcal{F}^*$, which follows in a similar way as in Lemma 3.28, and $\mu = \delta_{\tilde{x}} \in \mathcal{F}^*_{X_N, X_M^0}$. Then $\mu(Q - q) = 0$ and, again by Proposition 3.40, $|\lambda(Q - q)| \leq \|\lambda - \mu\|_{\mathcal{F}^*} \cdot \|Q\|_{\mathcal{F}}$. Now we proceed as in the proof of Theorem 3.32 and obtain (3.50), cf. (3.31). □

Theorems 3.35, 3.39 and 3.41 will be used in the next chapter to construct a Lyapunov function.

4

Construction of Lyapunov Functions

In Chapter 3 we have introduced radial basis functions. This method enables us to approximate a function via its values of a linear operator at a finite number of grid points. In particular, given the orbital derivative of a function Q on a grid, we can reconstruct the function Q by an approximation q. Moreover, we can estimate the error $|Q'(x) - q'(x)|$ and we have proven that it tends to zero if the fill distance of the grid tends to zero.

Hence, the approximations $q = t$ of T and $q = v$ of V via radial basis functions are functions with negative orbital derivatives themselves, provided that the grid is dense enough. For the example (2.11) the result is shown in Figures 4.1 and 4.2. We approximate the Lyapunov function V for which $V'(x) = -\|x-x_0\|^2$ holds by the approximant v. In Figure 4.1, left, the values of the approximating Lyapunov function $v(x)$ are shown, whereas in Figure 4.2 we show the approximation error $v'(x) - V'(x)$. In Figure 4.1, right, the sign of $v'(x)$ is given. Obviously, the approximating function v does not satisfy $v'(x) < 0$ for all $x \in K \setminus \{x_0\}$. There are points near x_0 where $v'(x) > 0$. One could think that the grid was not fine enough. But in fact the error estimate shows that one can only guarantee that $v'(x) < 0$ holds for a set $K \setminus U$, where U is an arbitrarily small neighborhood of x_0. Moreover, we show that in general there are points x near x_0 such that $v'(x) > 0$ holds, cf. Lemma 4.3. Hence, we call these approximating functions *non-local Lyapunov functions.* In order to obtain a Lyapunov function, i.e. a function with negative orbital derivative in $K \setminus \{x_0\}$, we have to combine the non-local, approximated Lyapunov function with some local information.

In Section 4.1 we use radial basis functions to approximate a global Lyapunov function of Section 2.3, i.e. T or V. Indeed, these functions have a certain smoothness and we know the values of their orbital derivative at all points, cf. Theorems 2.38 and 2.46. Hence, we can use an arbitrary grid. We obtain an approximation t of T, v of V, respectively. This approximation will itself have negative orbital derivative. However, the negativity of the orbital derivative cannot be guaranteed (i) near x_0 and (ii) near $\partial A(x_0)$, but only for each compact set $K \subset A(x_0) \setminus \{x_0\}$.

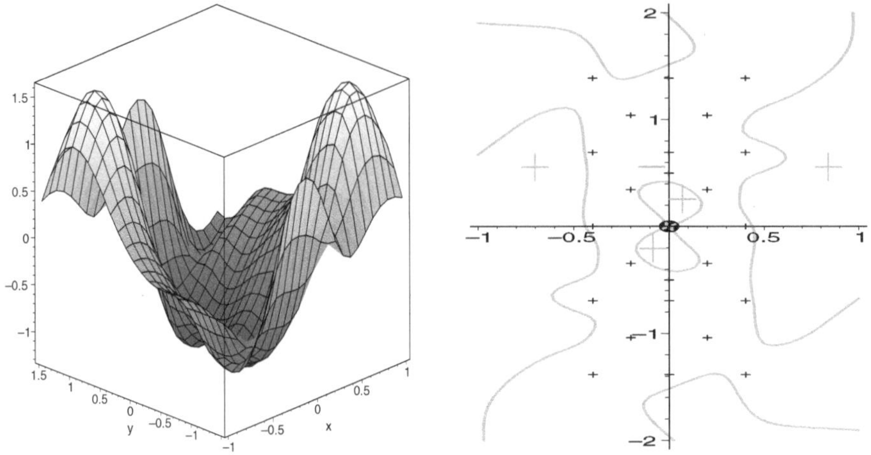

Fig. 4.1. Left: the approximating Lyapunov function $v(x, y)$. Right: the sign of $v'(x, y)$ (grey) and the grid (black +). v was obtained approximating the function $V(x, y)$ with $V'(x, y) = -(x^2 + y^2)$ for (2.11).

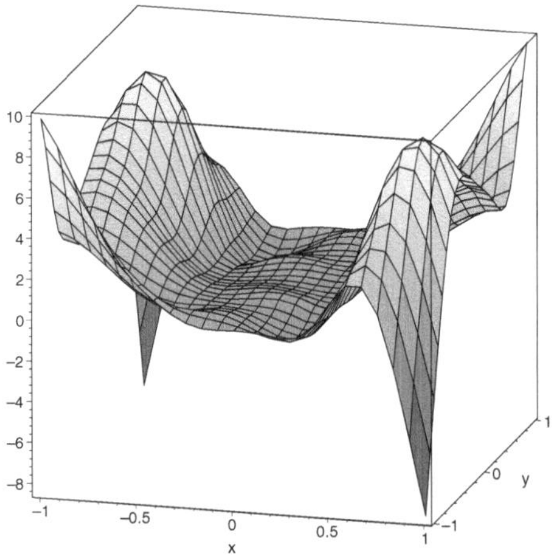

Fig. 4.2. The difference $v'(x, y) - V'(x, y)$. This is the approximation error, which is small in the region, where the grid points were set, and large outside.

In Section 4.2 we discuss three different possibilities to deal with the local problem (i) near x_0 since we can also show that in general there are points near x_0 with positive orbital derivative. For all three possibilities we use some local information: We determine a local Lyapunov basin, i.e. we calculate the local Lyapunov function $\mathfrak{d}$ or $\mathfrak{v}$ and find a Lyapunov basin for this Lyapunov function, and combine it with the approximation q of a global Lyapunov function. This can either be done using Theorem 2.26 or by construction of a new Lyapunov function combining the local and the approximated one by a partition of unity. The third possibility uses a Taylor polynomial of V to define a function W and then approximates W using radial basis functions.

The question (ii), whether we reach the whole basin of attraction or at least each compact subset with this method, will be addressed in Chapter 5.

4.1 Non-Local Part

In this section we prove that we can construct a function with negative orbital derivative in the non-local part, i.e. in a set which does not include a (local) neighborhood of x_0. The idea is to use one of the Lyapunov functions T or V outside a neighborhood of x_0 and to approximate it by radial basis functions via the orbital derivative. If the grid is dense enough, one obtains an approximating function with negative orbital derivative outside a neighborhood of the equilibrium. We make this precise using the estimates of Section 3.2.3.

We consider the Lyapunov functions T satisfying $T' = -\bar{c}$ and V satisfying $V' = -p(x)$. T is neither defined in x_0 nor outside $A(x_0)$. We extend T in a smooth way to a C^σ-function defined on $\mathbb{R}^n$ by setting $T(x) = 0$ for all x in a neighborhood U of x_0 and for all $x \notin A(x_0)$. Note that h^* in Theorem 4.1 thus depends on U, since C^* includes the factor $\|T\|_{\mathcal{F}}$ and the extension T depends on U. Moreover, in order to use the Wendland functions, T and hence f must be of a certain smoothness C^σ depending on the space dimension n and the parameter k of the Wendland function.

Theorem 4.1. *Consider the function $\Psi(x) = \psi_{l,k}(c\|x\|)$ with $c > 0$, where $\psi_{l,k}$ denotes the Wendland function with $k \in \mathbb{N}$ and $l := \left\lfloor \frac{n}{2} \right\rfloor + k + 1$. Let $f \in C^\sigma(\mathbb{R}^n, \mathbb{R}^n)$, where $\mathbb{N} \ni \sigma \geq \sigma^* := \frac{n+1}{2} + k$.*

Let $K_0 \subset A(x_0)$ be a compact set and U an open neighborhood of x_0 such that $K_0 \setminus U \neq \varnothing$. Let $X_N := \{x_1, \ldots, x_N\} \subset K_0 \setminus U$ be a grid with sufficiently small fill distance $h > 0$ in $K_0 \setminus U$.

Then there is a constant $C^* = C^*(K_0, U)$ such that for all reconstructions $t \in C^{2k-1}(\mathbb{R}^n, \mathbb{R})$ of T, cf. Theorem 2.38, with respect to the grid X_N and the operator $Dq(x) = q'(x)$ (orbital derivative), cf. Definition 3.4, we have

$$|t'(x) + \bar{c}| \leq C^* h^\kappa \qquad \text{holds for all } x \in K_0 \setminus U, \tag{4.1}$$

where $\kappa = \frac{1}{2}$ for $k = 1$ and $\kappa = 1$ for $k \geq 2$.

In particular, for all grids X_N with fill distance $h < h^* := \left(\frac{\bar{c}}{C^*}\right)^{\frac{1}{\kappa}}$ in $K_0 \setminus U$

$$t'(x) < 0 \qquad \text{holds for all } x \in K_0 \setminus U. \tag{4.2}$$

PROOF: Theorem 2.38 shows the existence of a function $T \in C^\sigma(A(x_0) \setminus \{x_0\}, \mathbb{R})$. With the open neighborhood $U \subset K_0$ of x_0 denote $K := K_0 \setminus U$.

We choose the Lyapunov function V_g of Theorem 2.46 for the differential equation $\dot{x} = g(x) = \frac{f(x)}{1+\|f(x)\|^2}$, cf. (2.2). Then $V_g \in C^\sigma(A(x_0), \mathbb{R})$ and the sets

$$K_1 := \{x \in A(x_0) \mid \min_{y \in K} V_g(y) \le V_g(x) \le \max_{y \in K} V_g(y)\} \supset K$$

$$\text{and } K_2 := \left\{x \in A(x_0) \mid \frac{1}{2} \min_{y \in K} V_g(y) \le V_g(x) \le \frac{3}{2} \max_{y \in K} V_g(y)\right\}$$

are compact sets by Theorem 2.46 with C^σ-boundary by Corollary 2.43. Hence, we can extend T to a function defined on $\mathbb{R}^n$ in a smooth way such that $T(x)$ remains unchanged on $K_1 \supset K$ and $T(x) = 0$ holds for all $x \notin K_2$. Denoting the new function also by T, we have $T \in C_0^\sigma(\mathbb{R}^n, \mathbb{R}) \subset \mathcal{F}$ by Lemma 3.13. The new function T also fulfills $T'(x) = -\bar{c}$ for all $x \in K$. Now we apply Theorem 3.35 to the function $Q = T$. Hence, (3.33) implies (4.1) if $h < H$, i.e. h is sufficiently small.

If the fill distance satisfies $h < h^* = \left(\frac{\bar{c}}{C^*}\right)^{\frac{1}{\kappa}}$, then (4.1) implies

$$t'(x) \le -\bar{c} + C^* h^\kappa < 0$$

for all $x \in K_0 \setminus U$, i.e. (4.2). □

In contrast to the function T, which is not defined in x_0, the function $V(x)$ is defined for all $x \in A(x_0)$. However, $V'(x_0) = 0$ so that we cannot guarantee $v'(x) < 0$ for all x since the error $|V'(x) - v'(x)|$ is possibly larger than $|V'(x)|$ near x_0. Thus, the grid constants h^* for T and V both depend on the neighborhood U. For V, the constant C^* only depends on K_0 and not on U.

Theorem 4.2. *Consider the function $\Psi(x) = \psi_{l,k}(c\|x\|)$ with $c > 0$, where $\psi_{l,k}$ denotes the Wendland function with $k \in \mathbb{N}$ and $l := \lfloor \frac{n}{2} \rfloor + k + 1$. Let $f \in C^\sigma(\mathbb{R}^n, \mathbb{R}^n)$, where $\mathbb{N} \ni \sigma \ge \sigma^* := \frac{n+1}{2} + k$. Let $K_0 \subset A(x_0)$ be a compact set. Let $X_N := \{x_1, \ldots, x_N\} \subset K_0 \setminus \{x_0\}$ be a grid with sufficiently small fill distance $h > 0$ in K_0.*

Then there is a constant $C^* = C^*(K_0)$ such that for all reconstructions $v \in C^{2k-1}(\mathbb{R}^n, \mathbb{R})$ of V with $V'(x) = -p(x)$, cf. Theorem 2.46, with respect to the grid X_N and the operator $Dq(x) = q'(x)$ (orbital derivative), cf. Definition 3.4, we have

$$|v'(x) + p(x)| \le C^* h^\kappa \qquad \text{holds for all } x \in K_0, \tag{4.3}$$

where $\kappa = \frac{1}{2}$ for $k = 1$ and $\kappa = 1$ for $k \ge 2$.

In particular, if U is an open neighborhood of x_0 with $K_0 \setminus U \neq \emptyset$, then for all grids $X_N \subset K_0 \setminus U$ with fill distance $h < h^* := \left(\frac{\min_{x \in K_0 \setminus U} p(x)}{C^*}\right)^{\frac{1}{\kappa}}$

$$v'(x) < 0 \qquad \text{holds for all } x \in K_0 \setminus U. \tag{4.4}$$

PROOF: Theorem 2.46 shows the existence of a function $V \in C^\sigma(A(x_0), \mathbb{R})$ with $V'(x) = -p(x)$ for all $x \in A(x_0)$. As in the proof of Theorem 4.1 denote by V_g the Lyapunov function of Theorem 2.46 for the differential equation $\dot{x} = g(x) = \frac{f(x)}{1+\|f(x)\|^2}$, cf. (2.2). Then $V_g \in C^\sigma(A(x_0), \mathbb{R})$ and the sets

$$K_1 := \{x \in A(x_0) \mid V_g(x) \leq \max_{y \in K_0} V_g(y)\} \supset K_0$$
$$\text{and } K_2 := \left\{x \in A(x_0) \mid V_g(x) \leq \frac{3}{2} \max_{y \in K_0} V_g(y)\right\}$$

are compact sets by Theorem 2.46 with C^σ-boundary by Proposition 2.42. Hence, we can extend V to a function defined on $\mathbb{R}^n$ in a smooth way such that $V(x)$ remains unchanged on $K_1 \supset K_0$ and $V(x) = 0$ holds for all $x \notin K_2$. Denoting the new function also by V, we have $V \in C_0^\sigma(\mathbb{R}^n, \mathbb{R}) \subset \mathcal{F}$ by Lemma 3.13. The new function V also fulfills $V'(x) = -p(x)$ for all $x \in K_0$. We apply Theorem 3.35 to the function $Q = V$ and the set K_0. Hence, (3.33) implies (4.3) if $h < H$, i.e. h is sufficiently small.

Now we choose a grid in $K := K_0 \setminus U$. If the fill distance of X_N in K satisfies $h < h^* = \left(\frac{\min_{x \in K_0 \setminus U} p(x)}{C^*}\right)^{\frac{1}{\kappa}}$, then (4.3) applied to K implies

$$v'(x) < -p(x) + \min_{x \in K_0 \setminus U} p(x) \leq 0$$

for all $x \in K_0 \setminus U$, i.e. (4.4). □

The estimates (4.2) and (4.4) show that $q'(x) < 0$ holds for all $x \in K_0 \setminus U$ where $q = t$ or $q = v$. As these results were obtained by inequalities, it is not clear whether they are sharp. Thus, we ask, whether in general there are points x near x_0 such that $q'(x) > 0$ holds. The following lemma gives a positive answer, if $\nabla q(x_0) \neq 0$. Since we do not impose any conditions on $\nabla q(x_0)$, in general $\nabla q(x_0) \neq 0$ holds and Lemma 4.3 implies that there is a direction $h \in \mathbb{R}^n$ such that $v'(x_0 + \delta h) > 0$ holds for all $\delta \in (0, \delta')$. On the other hand $q'(x_0 - \delta h) < 0$ holds for all $\delta \in (0, \delta')$. This can be observed in many examples, e.g. Figures 6.5, 6.6, 6.7, etc. However, in some examples, e.g. Figures 4.1 and 5.1 to 5.3 or Figure 6.12, one finds that $q'(x_0 + \delta h) > 0$ holds for $\delta \in (-\delta', 0) \cup (0, \delta')$. The reason for this behavior is that due to the symmetry of both the differential equation and the grid also the approximant q is symmetric and thus $\nabla q(x_0) = 0$, cf. Lemma 4.4. In this case one has to consider the higher order terms, cf. Lemma 4.5.

Lemma 4.3. *Consider $\dot{x} = f(x)$, $f \in C^1(\mathbb{R}^n, \mathbb{R}^n)$ and let x_0 be an equilibrium such that $-\nu < 0$ is the maximal real part of all eigenvalues of $Df(x_0)$. Let $q \in C^1(\mathbb{R}^n, \mathbb{R})$ be a function satisfying $\nabla q(x_0) \neq 0$. Then there is an $h \in \mathbb{R}^n$ and a $\delta' > 0$ such that*

$$q'(x_0 + \delta h) > 0 \textit{ holds for all } \delta \in (0, \delta') \tag{4.5}$$
$$\textit{and } q'(x_0 - \delta h) < 0 \textit{ holds for all } \delta \in (0, \delta'). \tag{4.6}$$

PROOF: Denote $w := \nabla q(x_0) \neq 0$. Let S be the matrix defined in Lemma 2.27 for $\epsilon = \frac{\nu}{2}$, where $-\nu < 0$ denotes the maximal real part of all eigenvalues of $Df(x_0) =: A$. Then by the same lemma $u^T B u \leq -\frac{\nu}{2}\|u\|^2$ holds for all $u \in \mathbb{R}^n$ with $B = SDf(x_0)S^{-1}$. Set

$$h := -S^{-1}(S^{-1})^T w \neq 0.$$

Then

$$\begin{aligned}
q'(x_0 + \delta h) &= \langle \nabla q(x_0 + \delta h), f(x_0 + \delta h)\rangle \\
&= \langle \nabla q(x_0), Df(x_0)\delta h\rangle + o(\delta) \\
&= \delta w^T Df(x_0) h + o(\delta) \\
&= -\delta w^T S^{-1} \underbrace{SDf(x_0)S^{-1}}_{=B}(S^{-1})^T w + o(\delta) \\
&= -\delta[(S^{-1})^T w]^T B (S^{-1})^T w + o(\delta)
\end{aligned}$$

for $\delta \to 0$. By Lemma 2.27 we have for $\delta > 0$: $q'(x_0 + \delta h) \geq \delta\frac{\nu}{2}\|(S^{-1})^T w\|^2 + o(\delta) > 0$ if $\delta \in (0, \delta')$ for some $\delta' > 0$. Similarly we obtain $q'(x_0 - \delta h) \leq -\delta\frac{\nu}{2}\|(S^{-1})^T w\|^2 + o(\delta) < 0$ if $\delta \in (0, \delta')$. □

Lemma 4.4 (Symmetry). *If both the dynamical system and the grid have the symmetry $f(x_0 - x) = -f(x_0 + x)$ and $X_{2N} = \{x_0 + x_1, \ldots, x_0 + x_N, x_0 - x_1, \ldots, x_0 - x_N\}$, then the reconstruction $q \in C^1(\mathbb{R}^n, \mathbb{R})$ of $T'(x) = -\bar{c}$ or $V'(x) = -p(x)$ with $p(x_0 - x) = p(x_0 + x)$ satisfies $q(x_0 - x) = q(x_0 + x)$ for all $x \in \mathbb{R}^n$. Thus, $\nabla q(x_0) = 0$.*

Proof: Denote the interpolation matrix $A = \begin{pmatrix} A^{++} & A^{+-} \\ A^{-+} & A^{--} \end{pmatrix}$ and $\alpha = \begin{pmatrix} \alpha^+ \\ \alpha^- \end{pmatrix}$, where A^{++}, A^{+-}, A^{-+} and A^{--} are $(N \times N)$ matrices and $\alpha^+, \alpha^- \in \mathbb{R}^N$. Denote the solution $\beta = \begin{pmatrix} \beta^+ \\ \beta^- \end{pmatrix}$, where $\beta^+, \beta^- \in \mathbb{R}^N$. Hence, the linear equation $A\beta = \alpha$ becomes

$$\begin{pmatrix} A^{++} & A^{+-} \\ A^{-+} & A^{--} \end{pmatrix} \begin{pmatrix} \beta^+ \\ \beta^- \end{pmatrix} = \begin{pmatrix} \alpha^+ \\ \alpha^- \end{pmatrix}. \tag{4.7}$$

Since $A = A^T$, we have $A^{+-} = A^{-+}$. We show $A^{++} = A^{--}$. Indeed, by (3.9) we have

$$\begin{aligned}
&A^{++}_{jk} \\
&= (\delta_{x_0+x_j} \circ D)^x (\delta_{x_0+x_k} \circ D)^y \Psi(x - y) \\
&= \psi_2(\|(x_0 + x_j) - (x_0 + x_k)\|)\langle x_j - x_k, f(x_0 + x_j)\rangle\langle x_k - x_j, f(x_0 + x_k)\rangle \\
&\quad -\psi_1(\|(x_0 + x_j) - (x_0 + x_k)\|)\langle f(x_0 + x_j), f(x_0 + x_k)\rangle \\
&= \psi_2(\|(x_0 - x_j) - (x_0 - x_k)\|)\langle -x_j + x_k, f(x_0 - x_j)\rangle\langle -x_k + x_j, f(x_0 - x_k)\rangle \\
&\quad -\psi_1(\|(x_0 - x_j) - (x_0 - x_k)\|)\langle f(x_0 - x_j), f(x_0 - x_k)\rangle \\
&= (\delta_{x_0-x_j} \circ D)^x (\delta_{x_0-x_k} \circ D)^y \Psi(x - y) \\
&= A^{--}_{jk}
\end{aligned}$$

using $f(x_0 - x) = -f(x_0 + x)$.

Now we show $\alpha^+ = \alpha^-$ for both $Q = T$ and $Q = V$. Indeed, by $p(x_0 - x) = p(x_0 + x)$ we have

$$\alpha_i^+ = Q'(x_0 + x_i) = \begin{cases} -\bar{c} \\ -p(x_0 + x_i) \end{cases} = \begin{cases} -\bar{c} \\ -p(x_0 - x_i) \end{cases} = Q'(x_0 - x_i) = \alpha_i^- .$$

Exchanging the first N and the last N equations in (4.7), we obtain

$$\begin{pmatrix} A^{--} & A^{-+} \\ A^{+-} & A^{++} \end{pmatrix} \begin{pmatrix} \beta^- \\ \beta^+ \end{pmatrix} = \begin{pmatrix} \alpha^- \\ \alpha^+ \end{pmatrix}.$$

Using $A^{++} = A^{--}$, $A^{+-} = A^{-+}$ and $\alpha^+ = \alpha^-$ this equation reads

$$\begin{pmatrix} A^{++} & A^{+-} \\ A^{-+} & A^{--} \end{pmatrix} \begin{pmatrix} \beta^- \\ \beta^+ \end{pmatrix} = \begin{pmatrix} \alpha^+ \\ \alpha^- \end{pmatrix}.$$

Since this system, which has the same matrix and right-hand side as (4.7) has a unique solution, $\beta^+ = \beta^- =: \beta$.

The approximation thus reads, cf. Proposition 3.5

$$\begin{aligned} q(x) &= \sum_{k=1}^{N} \beta_k \langle x_0 + x_k - x, f(x_0 + x_k) \rangle \psi_1(\|x - (x_0 + x_k)\|) \\ &\quad + \sum_{k=1}^{N} \beta_k \langle x_0 - x_k - x, f(x_0 - x_k) \rangle \psi_1(\|x - (x_0 - x_k)\|) \end{aligned}$$

Hence, for all $x \in \mathbb{R}^n$ we have, using $f(x_0 - x) = -f(x_0 + x)$

$$\begin{aligned} q(x_0 + x) &= \sum_{k=1}^{N} \beta_k \langle x_k - x, f(x_0 + x_k) \rangle \psi_1(\|x - x_k\|) \\ &\quad + \sum_{k=1}^{N} \beta_k \langle -x_k - x, f(x_0 - x_k) \rangle \psi_1(\|x + x_k\|) \\ &= \sum_{k=1}^{N} \beta_k \langle x - x_k, f(x_0 - x_k) \rangle \psi_1(\|x_k - x\|) \\ &\quad + \sum_{k=1}^{N} \beta_k \langle x_k + x, f(x_0 + x_k) \rangle \psi_1(\| - x - x_k\|) \\ &= q(x_0 - x) \end{aligned}$$

□

This symmetry occurs in the examples (2.11) and (6.4). If we use a symmetric grid, then the approximating functions are also symmetric. In these cases the assumptions of Lemma 4.3 are not fulfilled and the Hessian matrix $\operatorname{Hess} q(x_0)$ determines the behavior of $q'(x)$ near x_0, cf. Lemma 4.5.

Lemma 4.5. *Consider* $\dot{x} = f(x)$, $f \in C^1(\mathbb{R}^n, \mathbb{R}^n)$ *and let* x_0 *be an equilibrium such that* $-\nu < 0$ *is the maximal real part of all eigenvalues of* $Df(x_0)$. *Let* $q \in C^2(\mathbb{R}^n, \mathbb{R})$ *be a function satisfying* $\nabla q(x_0) = 0$.

If the symmetric matrix $\operatorname{Hess} q(x_0)$ *has at least one negative eigenvalue, then there is an* $h \in \mathbb{R}^n$ *and a* $\delta' > 0$ *such that*

$$q'(x_0 + \delta h) > 0 \text{ holds for all } \delta \in (-\delta', 0) \cup (0, \delta'). \tag{4.8}$$

PROOF: Since $\nabla q(x_0) = 0$, Taylor's Theorem yields

$$\nabla q(x) = \operatorname{Hess} q(x_0)(x - x_0) + o(\|x - x_0\|) \text{ for } \|x - x_0\| \to 0.$$

Since $\operatorname{Hess} q(x_0)$ is symmetric, we have for all $h \in \mathbb{R}^n$ and $\delta \in \mathbb{R}$

$$\begin{aligned} q'(x_0 + \delta h) &= \langle \nabla q(x_0 + \delta h), f(x_0 + \delta h) \rangle \\ &= \langle \operatorname{Hess} q(x_0) \delta h, Df(x_0) \delta h \rangle + o(\delta^2) \\ &= \delta^2 h^T \operatorname{Hess} q(x_0) Df(x_0) h + o(\delta^2) \\ &= \frac{\delta^2}{2} h^T \left[\operatorname{Hess} q(x_0) Df(x_0) + Df(x_0)^T \operatorname{Hess} q(x_0) \right] h + o(\delta^2) \end{aligned}$$

for $\delta \to 0$. Thus, the sign of $q'(x_0 + \delta h)$ for small $|\delta|$ is determined by the sign of $h^T \left[\operatorname{Hess} q(x_0) Df(x_0) + Df(x_0)^T \operatorname{Hess} q(x_0) \right] h$.

Denote the symmetric matrix

$$\operatorname{Hess} q(x_0) Df(x_0) + Df(x_0)^T \operatorname{Hess} q(x_0) =: -C. \tag{4.9}$$

We study the problem from a different point of view: Given a positive definite matrix C, by Theorem 2.30 there is one and only one solution $\operatorname{Hess} q(x_0)$ of (4.9) and $\operatorname{Hess} q(x_0)$ is positive definite. In a similar way one can prove that given a positive semidefinite matrix C, the unique solution $\operatorname{Hess} q(x_0)$ of (4.9) by Theorem 2.30 is positive semidefinite. Thus, if $\operatorname{Hess} q(x_0)$ has at least one negative eigenvalue, which is equivalent to $\operatorname{Hess} q(x_0)$ not being positive semidefinite, then also C is not positive semidefinite, i.e. there is an $h \in \mathbb{R}^n$ such that $h^T C h < 0$. □

Thus, the approximants t and v of Theorem 4.1 and Theorem 4.2 generally have no negative orbital derivative for all points near x_0 and, hence, are no Lyapunov functions themselves. Near x_0, however, we can use the local information provided by the linearization of f. In the following Section 4.2 we discuss different methods to deal with the local part of the problem.

4.2 Local Part

For the approximation of both T and V we have derived error estimates which guarantee that the approximations have negative orbital derivatives except for a neighborhood of x_0. We have also shown that in general the approximations indeed have points in each neighborhood of x_0 where the orbital derivative is positive. Hence, we have to deal with the local part. However, the local part is easily accessible since the linearization at x_0 provides local information. We use this local information in different ways and thus discuss three possibilities to deal with local part: in Section 4.2.1 we use a local Lyapunov basin to cover the set with positive orbital derivative. In Section 4.2.2 we use a local Lyapunov function and define a new function combining the local Lyapunov function and the calculated non-local Lyapunov function. This combined function turns out to be a Lyapunov function. In Section 4.2.3 we use the Taylor polynomial of V and approximate the function $W = \frac{V}{\mathfrak{n}}$ instead of V.

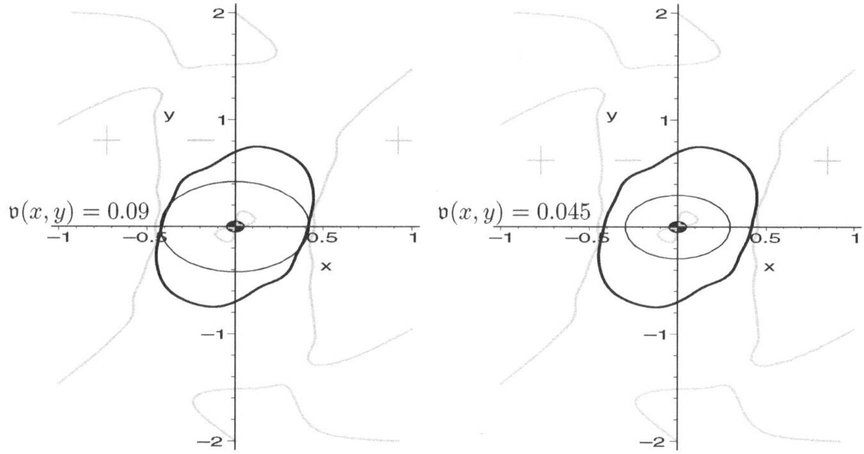

Fig. 4.3. v is an approximation of the Lyapunov function V using a hexagonal grid with grid distance $\alpha = 0.2$. The figures show the sign of $v'(x, y)$ (grey), a level set of v (black) and two different local Lyapunov basins of the local Lyapunov function $\mathfrak{v}$ (thin black). Left: $\mathfrak{v}(x, y) \leq 0.09$. Right: $\mathfrak{v}(x, y) \leq 0.045$. In the left-hand case, the local Lyapunov basin is not a subset of the Lyapunov basin, in the right-hand case it is. Hence, in the right-hand case one can define a Lyapunov function which is a combination of the local and the approximated one, using a partition of unity. This is not possible in the left-hand case. However, in both cases the Lyapunov basin (black) is a subset of the basin of attraction by Theorem 2.26. The example considered is (2.11).

Let us discuss the three possibilities applied to the example (2.11):

(i) The local information can be obtained by studying the linearization, namely the Jacobian $Df(x_0)$. This information can be used in form of a local Lyapunov basin. If the points with $q'(x) > 0$ near x_0 belong to a local Lyapunov basin, then Theorem 2.26 on Lyapunov basins with exceptional set implies that the largest sublevel set $q(x) \leq (R^*)^2$ such that for all points either $q'(x) < 0$ holds or x belongs to the local Lyapunov basin is a subset of the basin of attraction. In Figure 4.3, left and right, one can use this result to show that the Lyapunov basin bounded by the level set of the approximated Lyapunov function v in black is a subset of the basin of attraction.

(ii) If the local Lyapunov basin is a proper subset of the Lyapunov basin obtained by the approximated function, then one can also define a Lyapunov function $q^*(x)$ satisfying $(q^*)'(x) < 0$ for all $x \in K \setminus \{x_0\}$. The function q^* is obtained by the local Lyapunov function $\mathfrak{v}$ and the non-local one q by a partition of unity. Since level sets of q outside the local Lyapunov basin are level sets of q^* there is no need to explicitly calculate q^*: using Theorem 2.24 on Lyapunov basins one can directly prove that the Lyapunov basin with Lyapunov function q^* is a subset of the basin of attraction. In Figure 4.3, right, this is possible, whereas in Figure 4.3, left, the local Lyapunov basin is not a subset of the Lyapunov basin and thus the construction is not possible.

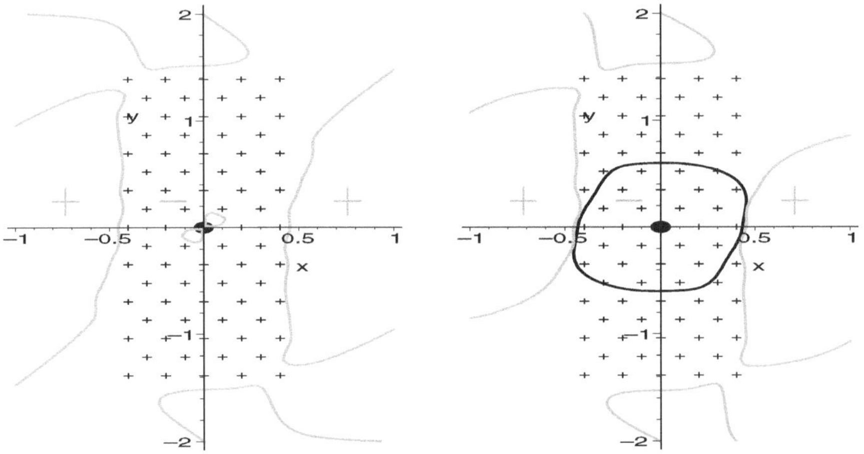

Fig. 4.4. Left: the direct approximation of V by v with a hexagonal grid (black +) of grid distance $\alpha = 0.2$. The sign of $v'(x, y)$ (grey) is positive for some points near 0. Right: the approximation of V via W using the Taylor polynomial with the same grid (black +). The sign of $v_W'(x, y)$ (grey) is negative for all points near 0. The level set $v_W(x, y) = 0.2$ (black) is shown which is the boundary of a Lyapunov basin. The Lyapunov basin is thus a subset of the basin of attraction for example (2.11).

(iii) The third approach to use the local information is only possible for $Q = V$ with $V'(x) = -\|x - x_0\|^2$: one calculates the Taylor polynomial $\mathfrak{h}(x)$ of $V(x)$ of a certain order and adds another high order polynomial to obtain a polynomial $\mathfrak{n}(x)$ for which $\mathfrak{n}(x) > 0$ holds for all $x \neq x_0$. We define the function $W(x) = \frac{V(x)}{\mathfrak{n}(x)}$ which satisfies a certain partial differential equation. We approximate the function W by w and define an approximation $v_W(x) = w(x)\mathfrak{n}(x)$ of $V(x)$; this approximation v_W, in contrast to the direct approximation v of V, has negative orbital derivative in $K \setminus \{x_0\}$ if the grid is fine enough and, thus, we do not need a local Lyapunov function. We apply this procedure to the example (2.11). The Taylor polynomial for V of order five reads, cf. Example 2.61:

$$\mathfrak{h}(x,y) = \frac{1}{2}x^2 + \frac{1}{2}y^2 + x^4 - \frac{3}{2}x^3y + \frac{11}{16}x^2y^2 + \frac{1}{32}xy^3 + \frac{3}{32}y^4.$$

In this case $\mathfrak{n}(x) = \mathfrak{h}(x) > 0$ holds for all $x \neq x_0$. The level set of the approximating function v_W and the sign of v'_W are shown in Figure 4.4, right. Note that the direct approximation v of V in general has points near x_0 where $v'(x) > 0$, cf. Figure 4.4, left, where the same grid was used.

4.2.1 Local Lyapunov Basin

Since we can obtain a local Lyapunov basin by one of the local Lyapunov functions $\mathfrak{q} = \mathfrak{d}$ or $\mathfrak{q} = \mathfrak{v}$, we only need to combine this local information with a non-local Lyapunov function. This is done in the following lemma, which also holds for arbitrary Lyapunov functions $\mathfrak{q}$ with Lyapunov basin $\tilde{K}$.

Lemma 4.6. *Let $\mathfrak{q}$ be a Lyapunov function with Lyapunov basin $\tilde{K} = \tilde{K}_r^{\mathfrak{q}}(x_0)$. Let $U := \tilde{B}_r^{\mathfrak{q}}(x_0)$.*

Consider the function $\Psi(x) = \psi_{l,k}(c\|x\|)$ with $c > 0$, where $\psi_{l,k}$ denotes the Wendland function with $k \in \mathbb{N}$ and $l := \lfloor \frac{n}{2} \rfloor + k + 1$. Let $f \in C^\sigma(\mathbb{R}^n, \mathbb{R}^n)$, where $\mathbb{N} \ni \sigma \geq \sigma^* := \frac{n+1}{2} + k$. Let either 1. or 2. hold:

1. *Let $A(x_0) \supset K_0$ be a compact set with $K_0 \setminus U \neq \varnothing$ and $X_N \subset K_0 \setminus U$ a grid with fill distance $h < h^*$, where h^* was defined in Theorem 4.1. Let $t \in C^{2k-1}(\mathbb{R}^n, \mathbb{R})$ be the reconstruction of T, cf. Theorem 2.38, with respect to the grid X_N with operator $Dq(x) = q'(x)$ (orbital derivative).*
2. *Let $A(x_0) \supset K_0$ be a compact set with $K_0 \setminus U \neq \varnothing$ and $X_N \subset K_0 \setminus U$ a grid with fill distance $h < h^*$, where h^* was defined in Theorem 4.2. Let $v \in C^{2k-1}(\mathbb{R}^n, \mathbb{R})$ be the reconstruction of V, cf. Theorem 2.46, with respect to the grid X_N with operator $Dq(x) = q'(x)$ (orbital derivative).*

Let K be a compact set and B be a neighborhood of K such that $K = \{x \in B \mid q(x) \leq R\} \subset K_0$, where $q = t$ or $q = v$, respectively, and $x_0 \in \mathring{K}$.

Then q is a Lyapunov function with Lyapunov basin K and exceptional set U in the sense of Definition 2.25.

PROOF: The properties in Definition 2.25 are fulfilled: 1. and 3. by assumption, and 2. because of (4.2) or (4.4). Finally, 4., i.e. $U \subset A(x_0)$ holds since $\overline{U}$ is a Lyapunov basin. □

4.2.2 Local Lyapunov Function

In this section we use a local Lyapunov function in order to modify the non-local one near the equilibrium; level sets of the non-local Lyapunov function keep being level sets of the modified one. Hence, there is no need to explicitly calculate the modified function in order to obtain a Lyapunov basin.

Assume that $\mathfrak{q}$ is a Lyapunov function with Lyapunov basin $\tilde{K} = \tilde{K}_r^{\mathfrak{q}}(x_0)$ in the sense of Definition 2.23. We also assume that $\mathfrak{q}$ is defined in $\mathbb{R}^n$. The functions $\mathfrak{d}$ and $\mathfrak{v}$ with Lyapunov basins are examples for such functions $\mathfrak{q}$, cf. Corollaries 2.29 and 2.33.

We define a function h which is zero near x_0 and one outside $\tilde{K}$ to link the local and the global part together. The construction of h in Lemma 4.7 is similar to a partition of unity. Note that we will prove $h'(x) \leq 0$ for its orbital derivative.

Lemma 4.7. *Let $\mathfrak{q} \in C^1(\mathbb{R}^n, \mathbb{R})$ be a Lyapunov function with Lyapunov basin $\tilde{K} = \tilde{K}_r^{\mathfrak{q}}(x_0)$. Let $0 < r_0 < r$. Then there is a function $h \in C^\infty(\mathbb{R}^n, \mathbb{R})$, such that*

- *$h(x) \in [0, 1]$ for all $x \in \mathbb{R}^n$,*
- *$h(x) = 0$ for $x \in \tilde{K}_{r-r_0}^{\mathfrak{q}}(x_0)$,*
- *$h(x) = 1$ for $x \notin \tilde{B}_r^{\mathfrak{q}}(x_0)$,*
- *$h'(x) \leq 0$ for all $x \in \mathbb{R}^n$,*
- *$h'(x) = 0$ for all $x \in \tilde{K}_{r-r_0}^{\mathfrak{q}}(x_0)$ and for all $x \notin \tilde{B}_r^{\mathfrak{q}}(x_0)$.*

PROOF: Following the standard procedure for a partition of unity, we define H_1 for all $\rho \in \mathbb{R}_0^+$:

$$H_1(\rho) := \begin{cases} \exp\left(\frac{1}{[r^2-\rho][(r-r_0)^2-\rho]}\right) & \text{for } (r-r_0)^2 < \rho < r^2 \\ 0 & \text{otherwise.} \end{cases}$$

H_1 is a C^∞-function and $H_1(\rho) \geq 0$ holds for all $\rho \in \mathbb{R}_0^+$. We set

$$H_2(\rho) := \frac{\int_{(r-r_0)^2}^{\rho} H_1(\tilde{\rho})\, d\tilde{\rho}}{\int_{(r-r_0)^2}^{r^2} H_1(\tilde{\rho})\, d\tilde{\rho}}.$$

$H_2(\rho)$ is monotonically increasing with respect to ρ. For $\rho \in \left[0, (r-r_0)^2\right]$, $H_2(\rho) = 0$ and for $\rho \in \left[r^2, \infty\right)$, $H_2(\rho) = 1$. Now set $h(x) := H_2(\mathfrak{q}(x))$ for $x \in \tilde{K}_r^{\mathfrak{q}}(x_0)$ and $h(x) = 1$ otherwise. We have $h \in C^\infty(\mathbb{R}^n, \mathbb{R})$ and the properties of h follow easily by the respective properties of H_2 and $\mathfrak{q}$, e.g. the orbital derivative $h'(x)$ for $x \in \tilde{B}_r^{\mathfrak{q}}(x_0)$ is given by $h'(x) = \underbrace{\frac{d}{d\rho} H_2(\mathfrak{q}(x))}_{\geq 0} \cdot \underbrace{\mathfrak{q}'(x)}_{\leq 0}$. □

In the next theorem we combine a (local) Lyapunov function $\mathfrak{q}$ with a non-local Lyapunov function q.

Theorem 4.8 (Extension). *Let $\mathfrak{q} \in C^s(\mathbb{R}^n, \mathbb{R})$, $s \geq 1$, be a Lyapunov function with Lyapunov basin $\tilde{K} = \tilde{K}_r^{\mathfrak{q}}(x_0)$.*

Let K be a compact set with $\tilde{B}_r^{\mathfrak{q}}(x_0) \subset K$. Let $q \in C^s(\mathbb{R}^n \setminus \{x_0\}, \mathbb{R})$ be a function such that

$$q'(x) < 0 \qquad \text{holds for all } x \in K \setminus \tilde{B}_r^{\mathfrak{q}}(x_0). \tag{4.10}$$

Then there is a function $q^* \in C^s(\mathbb{R}^n, \mathbb{R})$, such that

- *$q^*(x) = aq(x) + b$ holds for all $x \notin \tilde{B}_r^{\mathfrak{q}}(x_0)$ with constants $a \geq 1$, $b > 0$,*
- *$(q^*)'(x) < 0$ holds for all $x \in K \setminus \{x_0\}$.*

Moreover, $q^*(x_0) = \mathfrak{q}(x_0)$ holds.

Proof: By the continuity of q and by (4.10) there is an $0 < r_0 < r$ such that

$$q'(x) < 0 \text{ holds for all } x \in K \setminus \tilde{B}_{r-r_0}^{\mathfrak{q}}(x_0). \tag{4.11}$$

Since $q(x)$ and $\mathfrak{q}(x)$ are continuous functions and since K is compact, the following maxima exist:

$$b_q := \max_{x \in K} |q(x)|, \tag{4.12}$$
$$b_{\mathfrak{q}} := \max_{x \in K} |\mathfrak{q}(x)|. \tag{4.13}$$

Since also $q'(x)$ and $\mathfrak{q}'(x)$ are continuous functions, the following maxima exist:

$$a_q := \max_{x \in \overline{\tilde{B}_r^{\mathfrak{q}}(x_0) \setminus \tilde{B}_{r-r_0}^{\mathfrak{q}}(x_0)}} q'(x) < 0, \tag{4.14}$$
$$a_{\mathfrak{q}} := \max_{x \in \overline{\tilde{B}_r^{\mathfrak{q}}(x_0) \setminus \tilde{B}_{r-r_0}^{\mathfrak{q}}(x_0)}} \mathfrak{q}'(x) < 0. \tag{4.15}$$

Set $a := \max\left(\frac{a_{\mathfrak{q}}}{a_q}, 1\right) \geq 1$ and $b := ab_q + b_{\mathfrak{q}} > 0$. Thus, $aq(x) - \mathfrak{q}(x) + b \geq 0$ holds for all $x \in K$. With the function h, cf. Lemma 4.7, we set

$$q^*(x) := h(x)[aq(x) + b] + [1 - h(x)]\mathfrak{q}(x).$$

Obviously, the function q^* is C^s since $h(x) = 0$ holds for all $x \in \tilde{K}_{r-r_0}^{\mathfrak{q}}(x_0)$. Note that $q^*(x_0) = \mathfrak{q}(x_0)$ holds. We show the properties of q^*. Since $h(x) = 1$ for all $x \notin \tilde{B}_r^{\mathfrak{q}}(x_0)$, the first property is clear.

We calculate the orbital derivative of q^*

$$(q^*)'(x) = h'(x)[aq(x) - \mathfrak{q}(x) + b] + h(x)aq'(x) + [1 - h(x)]\mathfrak{q}'(x). \tag{4.16}$$

For $x \in K \setminus \{x_0\}$ we distinguish between the three cases $x \notin \tilde{B}_r^{\mathfrak{q}}(x_0)$, $\sqrt{\mathfrak{q}(x)} \in (r - r_0, r)$ and $0 < \sqrt{\mathfrak{q}(x)} \leq r - r_0$.

Case 1: $x \notin \tilde{B}_r^{\mathfrak{q}}(x_0)$
Here we have $h(x) = 1$, $h'(x) = 0$ and $q'(x) < 0$ by (4.11). Thus, with (4.16) we have

$$(q^*)'(x) = \underbrace{h'(x)}_{=0}[aq(x) - \mathfrak{q}(x) + b] + \underbrace{h(x)}_{=1} aq'(x) + \underbrace{[1 - h(x)]}_{=0} \mathfrak{q}'(x) < 0.$$

Case 2: $\sqrt{\mathfrak{q}(x)} \in (r - r_0, r)$
Here we have $h(x) \in [0, 1]$, $h'(x) \leq 0$, cf. Lemma 4.7, $q'(x) < 0$ by (4.11) and $\mathfrak{q}'(x) < 0$. Hence, with (4.16) we have

$$\begin{aligned}(q^*)'(x) &= \underbrace{h'(x)}_{\leq 0}\underbrace{[aq(x) - \mathfrak{q}(x) + b]}_{\geq 0} + h(x)aq'(x) + [1 - h(x)]\mathfrak{q}'(x) \\ &\leq 0 + h(x)aa_q + a_{\mathfrak{q}}[1 - h(x)] \text{ by (4.14) and (4.15)} \\ &= a_{\mathfrak{q}} + h(x)\underbrace{(aa_q - a_{\mathfrak{q}})}_{\leq 0} \\ &= a_{\mathfrak{q}} < 0.\end{aligned}$$

Case 3: $0 < \sqrt{\mathfrak{q}(x)} \leq r - r_0$
Here we have $h(x) = h'(x) = 0$ and $\mathfrak{q}'(x) < 0$. Hence, $(q^*)'(x) = \mathfrak{q}'(x) < 0$. □

Applying the Extension Theorem 4.8 to a pair of local and non-local Lyapunov function $\mathfrak{q}$ and q we obtain the following corollary which proves the existence of a function q^* with negative orbital derivatives also near x_0. Moreover, since level sets of q away from x_0 are level sets of q^*, one can find a Lyapunov basin without explicitly calculating q^*.

Corollary 4.9 *Let $\mathfrak{q} \in C^s(\mathbb{R}^n, \mathbb{R})$, $s \geq 1$, be a Lyapunov function with Lyapunov basin $\tilde{K} = \tilde{K}_r^{\mathfrak{q}}(x_0)$. Let $q \in C^s(\mathbb{R}^n, \mathbb{R})$, B be an open set and $\tilde{B}_r^{\mathfrak{q}}(x_0) \subset K \subset B$ be a compact set such that*

$$q'(x) < 0 \qquad \text{holds for all } x \in K \setminus \tilde{B}_r^{\mathfrak{q}}(x_0) \tag{4.17}$$

and $K = \{x \in B \mid q(x) \leq R^2\}$. Then $K \subset A(x_0)$ and K is positively invariant.

Proof: The Extension Theorem 4.8 yields the existence of a function q^*. We claim that q^* is a Lyapunov function with Lyapunov basin $K = \{x \in B \mid q^*(x) \leq (R^*)^2\}$, where $R^* := \sqrt{aR^2 + b}$. Then the corollary follows by Theorem 2.24.

Indeed, we only have to show that for all $x \in \tilde{B}_r^{\mathfrak{q}}(x_0)$ the inequality $q^*(x) < (R^*)^2$ holds. Let $x \in \tilde{B}_r^{\mathfrak{q}}(x_0)$. Since $\tilde{K}_r^{\mathfrak{q}}(x_0)$ is a Lyapunov basin, $x \in A(x_0)$. Since $\Omega = \partial\tilde{K}_r^{\mathfrak{q}}(x_0)$ is a non-characteristic hypersurface by Lemma 2.37, there is a time $t < 0$, such that $\mathfrak{q}(S_t x) = r^2$. Since $q^*(y) \leq (R^*)^2$ holds for all $y \in \partial\tilde{K}_r^{\mathfrak{q}}(x_0) \subset K$, we have

$$0 < \int_0^t (q^*)'(S_\tau x)\, d\tau = q^*(S_t x) - q^*(x) \leq (R^*)^2 - q^*(x),$$

which proves $q^*(x) < (R^*)^2$. □

The difference between Sections 4.2.1 and 4.2.2 is the way we dealt with local part: in Section 4.2.2 we defined a new Lyapunov function whereas in Section 4.2.1 we used Theorem 2.26 with an exceptional set. In Section 4.2.2 we need $\tilde{B}_r^{\mathfrak{q}}(x_0) \subset K$, can apply the results of Section 2.3.2 and we obtain that K is positively invariant. In Section 4.2.1 there may be points in $x \in \tilde{B}_r^{\mathfrak{q}}(x_0) \setminus K$ and, thus, we do not obtain the positive invariance of K. However, the set $\tilde{B}_r^{\mathfrak{q}}(x_0) \cup K$ is positively invariant: $\tilde{B}_r^{\mathfrak{q}}(x_0)$ is positively invariant by definition. Thus, an orbit can leave $\tilde{B}_r^{\mathfrak{q}}(x_0) \cup K$ only through a point $w \in \partial K \setminus \tilde{B}_r^{\mathfrak{q}}(x_0)$. This is not possible by the same argument as in Theorem 2.26.

The application of the results of Sections 4.2.1 and 4.2.2 to several examples is discussed in Section 6.1.

4.2.3 Taylor Polynomial

In this section we consider a different way to obtain a Lyapunov function v_W via approximation. This function v_W fulfills $v_W'(x) < 0$ for all x near x_0. The function V, satisfying $V'(x) = -\|x-x_0\|^2$ when directly approximated, results in an approximation v which has positive orbital derivative near x_0 in general. The reason for this fact is that we obtain the error estimate $|V'(x) - v'(x)| \leq \iota := C^* h^\kappa$ with a fixed ι for a certain fill distance h of the grid, and thus $v'(x) \leq -\|x - x_0\|^2 + \iota$ which is only negative for $\|x - x_0\|^2 > \iota$.

In this section we consider again the function V satisfying $V'(x) = -\|x - x_0\|^2$. We use its Taylor polynomial of order P and the function $\mathfrak{n}$ defined in Definition 2.56 and write $V(x) = \mathfrak{n}(x)W(x)$. Instead of approximating V directly, we approximate the function W by an approximation w and thus obtain an approximation $v_W(x) = \mathfrak{n}(x)w(x)$ of V. Recall that $\mathfrak{n}$ is a known function satisfying $\mathfrak{n}(x) > 0$ for $x \neq x_0$. The function W is a C^{P-2}-function, cf. Proposition 2.58.

Theorem 4.10. *Consider the function $\Psi(x) = \psi_{l,k}(c\|x\|)$ with $c > 0$, where $\psi_{l,k}$ denotes the Wendland function with $k \in \mathbb{N}$ and $l := \lfloor \frac{n}{2} \rfloor + k + 1$.*

Let $V(x)$ be the function such that $V'(x) = -\|x - x_0\|^2$ and $V(x_0) = 0$, and $\mathfrak{n}(x)$ as in Definition 2.56 with $P \geq 2 + \sigma^*$ where $\sigma^* := \frac{n+1}{2} + k$. Let $f \in C^\sigma(\mathbb{R}^n, \mathbb{R}^n)$ with $\sigma \geq P$.

Let $K \subset A(x_0)$ be a compact set with $x_0 \in \mathring{K}$. Let $X_N := \{x_1, \ldots, x_N\} \subset K \setminus \{x_0\}$ be a grid with $0 < \tilde{c} < \frac{1}{C}$, cf. Theorem 3.39 for the meaning of $\tilde{c}$, where C is defined in Proposition 2.58, 2. with respect to K.

Let $w \in C^{2k-1}(\mathbb{R}^n, \mathbb{R})$ be the reconstruction of $W = \frac{V}{\mathfrak{n}}$, cf. Proposition 2.58, with respect to the grid X_N with operator $D_m w(x) = w'(x) + m(x)w(x)$, where $m(x) = \frac{\mathfrak{n}'(x)}{\mathfrak{n}(x)}$. Set $v_W(x) = w(x)\mathfrak{n}(x)$.

Then we have

$$v_W'(x) < 0 \quad \text{for all } x \in K \setminus \{x_0\} \tag{4.18}$$

and, moreover, $$\left|v_W'(x) + \|x - x_0\|^2\right| \leq \tilde{c}C\|x - x_0\|^2. \tag{4.19}$$

PROOF: $W \in C^{P-2}(A(x_0), \mathbb{R})$ defined in Proposition 2.58 can be extended smoothly such that the new function coincides with the former one on K and has compact support, cf. the proof of Theorem 4.1. We denote the extension again by $W \in C_0^{P-2}(\mathbb{R}^n, \mathbb{R})$; since $P - 2 \geq \sigma^*$, we have $W \in \mathcal{F}$ by Lemma 3.13. By Theorem 3.39 the reconstruction w of W satisfies, cf. (3.42)

$$D_m W(x) - \tilde{c} \leq D_m w(x) \leq D_m W(x) + \tilde{c} \tag{4.20}$$

for all $x \in K \setminus \{x_0\}$. Since $V(x) = \mathfrak{n}(x)W(x)$ we have

$$-\|x - x_0\|^2 = V'(x) = \mathfrak{n}'(x)W(x) + \mathfrak{n}(x)W'(x).$$

For $x \in K \setminus \{x_0\}$ we can divide this equation by $\mathfrak{n}(x) \neq 0$ and obtain

$$W'(x) + \frac{\mathfrak{n}'(x)}{\mathfrak{n}(x)}W(x) = -\frac{\|x - x_0\|^2}{\mathfrak{n}(x)} \tag{4.21}$$

$$D_m W(x) = -\frac{\|x - x_0\|^2}{\mathfrak{n}(x)} \tag{4.22}$$

with $D_m W(x) = \langle \nabla W(x), f(x) \rangle + \frac{\mathfrak{n}'(x)}{\mathfrak{n}(x)} W(x) = W'(x) + m(x)W(x)$, where $m(x) := \frac{\mathfrak{n}'(x)}{\mathfrak{n}(x)}$. By Proposition 2.58, 2., we have

$$\mathfrak{n}(x) \leq C\|x - x_0\|^2 \tag{4.23}$$

for all $x \in K$.

Note that both $\frac{\mathfrak{n}'(x)}{\mathfrak{n}(x)} = m(x)$ and $\frac{\|x-x_0\|^2}{\mathfrak{n}(x)}$, the left- and the right-hand side of (4.21), are not defined in x_0 but bounded in each neighborhood of x_0, cf. Proposition 3.38. However, we can approximate the smooth function W by (4.22) outside x_0. For $v_W(x) = w(x)\mathfrak{n}(x)$ we have

$$\begin{aligned} v_W'(x) &= w'(x)\mathfrak{n}(x) + w(x)\mathfrak{n}'(x) \\ &= \mathfrak{n}(x) D_m w(x) \\ &\leq \mathfrak{n}(x) \left(D_m W(x) + \tilde{c}\right) \text{ by (4.20)} \\ &= -\|x - x_0\|^2 + \tilde{c}\,\mathfrak{n}(x) \text{ by (4.22)} \\ &\leq -\|x - x_0\|^2 (1 - \tilde{c}C) \text{ by (4.23)} \\ &< 0 \end{aligned} \tag{4.24}$$

for all $x \in K \setminus \{x_0\}$, since $\tilde{c} < \frac{1}{C}$. This shows (4.18).

Moreover, we have

$$\begin{aligned} v_W'(x) &= \mathfrak{n}(x) D_m w(x) \\ &\geq \mathfrak{n}(x) \left(D_m W(x) - \tilde{c}\right) \text{ by (4.20)} \\ &\geq -\|x - x_0\|^2 (1 + \tilde{c}C) \end{aligned} \tag{4.25}$$

for all $x \in K \setminus \{x_0\}$ by (4.22) and (4.23). Finally, (4.24) and (4.25) show (4.19), which proves the theorem. □

Examples for the approximation of V via the Taylor polynomial are given in Section 6.2.

5

Global Determination of the Basin of Attraction

Up to now we have focussed on the construction of a Lyapunov function, i.e. a function with negative orbital derivative. For such a function in a neighborhood of an exponentially asymptotically stable equilibrium one can always find a Lyapunov basin as described above. Since the basin of attraction is an open set and the Lyapunov basins are compact sets, they are always proper subsets of the basin of attraction. Hence, the best we can expect is that, given a compact subset of the basin of attraction, we find a Lyapunov basin larger than this compact set with our method.

For the results of this section we assume that f is bounded in $A(x_0)$ which can be achieved by studying the dynamically equivalent system $\dot{x} = g(x)$ where $g(x) = \frac{f(x)}{1+\|f(x)\|^2}$, cf. Remark 2.5. In particular, the basin of attraction of x_0 is the same for the two systems $\dot{x} = f(x)$ and $\dot{x} = g(x)$. We can show that given a compact set $K_0 \subset A(x_0)$ one obtains a Lyapunov basin larger than K_0 by approximating the function V. The approximation can either be direct or using the Taylor polynomial of V. This result uses an estimate of $|[V(x) - V(x_0)] - [v(x) - v(x_0)]|$ near x_0. Note, that this estimate is possible although the approximation v only uses the values of the orbital derivative $V'(x)$ and not of $V(x)$. The reason is that V is a smooth function in x_0. Thus, the result does not hold for approximations of the function T.

The result requires a sufficiently dense grid. Even if the set $v'(x) < 0$ is already quite large, the largest sublevel set of v probably only provides a small Lyapunov basin. In order to enlarge the Lyapunov basin one has to use a denser grid – not only where $v'(x) > 0$, not only near the boundary of the former Lyapunov basin, but in the whole expected basin of attraction.

We consider again the example (2.11). A series of Lyapunov basins with denser grids is shown in Figures 5.1 to 5.3. Note that here the sets $v'(x) < 0$ do not change significantly, but the values of v and hence the Lyapunov basins do. However, the enlargement of the Lyapunov basins is not monotonous, cf. Figures 5.1 and 5.2. This indicates that in practical applications the error is in fact smaller than predicted by the corresponding theorem. Figure 5.3

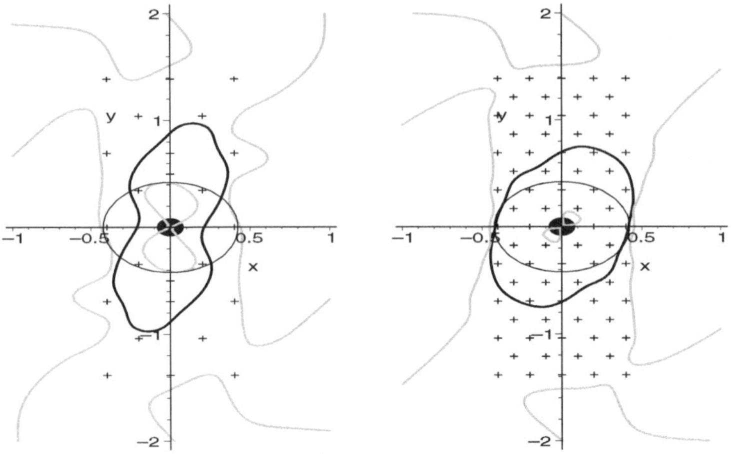

Fig. 5.1. The function v is the approximation of the Lyapunov function V satisfying $V'(x, y) = -(x^2 + y^2)$ for the example (2.11). The figures show the set $v'(x, y) = 0$ (grey), a level set of $v(x, y)$ (black) and a local Lyapunov basin (thin black). Left: grid density $\alpha = 0.4$. Right: grid density $\alpha = 0.2$ (cf. also Figures 5.2 and 5.3). Compare the Lyapunov basins (black) in both figures: although we used more grid points (black +) in the right-hand figure, the left-hand Lyapunov basin is no subset of the right-hand one.

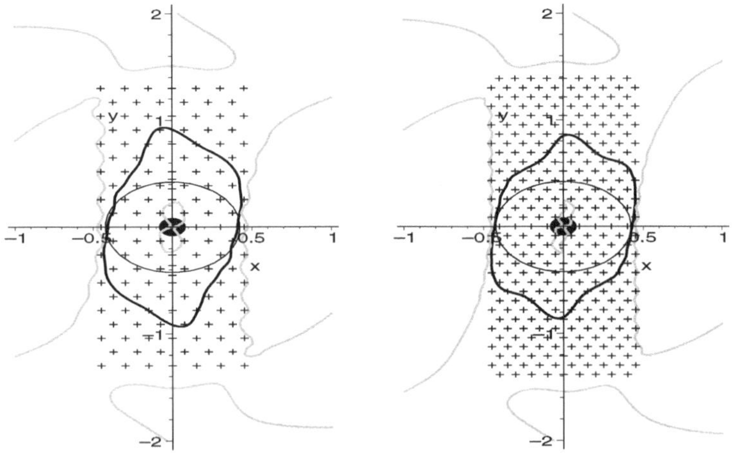

Fig. 5.2. The set $v'(x, y) = 0$ (grey), a level set of $v(x, y)$ (black), the grid points (black +) and a local Lyapunov basin (thin black). Left: grid density $\alpha = 0.15$. Right: grid density $\alpha = 0.1$ (cf. also Figure 5.3). The example considered is (2.11).

compares the best result (484 grid points) with the numerically calculated boundary of the basin of attraction, an unstable periodic orbit. For the data of the grids and the calculations of all figures, cf. Appendix B.2.

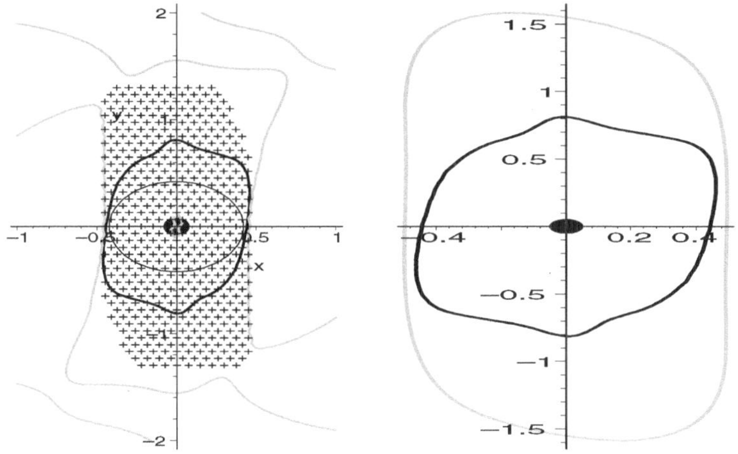

Fig. 5.3. The set $v'(x, y) = 0$ (grey), a level set of $v(x, y)$ (black), the grid points (black +) and a local Lyapunov basin (thin black). Left: grid density $\alpha = 0.075$. Right: the same Lyapunov basin (black) is shown together with the numerically calculated periodic orbit which is the boundary of the basin of attraction (grey). The example considered is (2.11).

Given a Lyapunov function q and a Lyapunov basin K, Theorem 2.24 implies $K \subset A(x_0)$. Thus, on the one hand, we search for a function q, the orbital derivative of which is negative in $K \setminus \{x_0\}$, i.e. a Lyapunov function. On the other hand, K is supposed to be a sublevel set of q, i.e. K is a Lyapunov basin. We have discussed the construction of a Lyapunov function in the preceding chapter, but can we thus find a Lyapunov basin? The appropriate question concerning a *global* Lyapunov function is, whether we can cover any compact set $K_0 \subset A(x_0)$ with our approach, supposed that the grid is dense enough.

The precise result which we will obtain reads: Let K_0 be a compact set with $x_0 \in \overset{\circ}{K}_0 \subset K_0 \subset A(x_0)$. Then there is an open set B with $\overline{B} \subset A(x_0)$, a compact set $B \supset K \supset K_0$ and a function q obtained by our method, such that

1. $K = \{x \in B \mid q(x) \leq (R^*)^2\}$ for an $R^* \geq 0$,
2. $q'(x) < 0$ holds for all $x \in K \setminus \{x_0\}$.

In other words, q is a Lyapunov function with Lyapunov basin K and thus they fulfill the conditions of Theorem 2.24. This can be achieved for the approximant v_W of V via W. For the direct approximation of V by v one has to use the extension v^* due to the local behavior of v. For the approximation t of T these results do not hold since T is not defined and smooth in x_0 which is necessary for the proof. For T, we consider a mixed approximation in Section 5.2.

We show in Section 5.1.1 that we can cover any compact subset of $A(x_0)$ when approximating the function V with $V' = -p(x)$. Because of possible numerical problems near x_0 – we have to choose a very dense grid here and this leads to a high condition number of the interpolation matrix – we also discuss a mixed approximation of V^* in Section 5.2 where, additionally to the orbital derivative, the values of the approximated function are given on a non-characteristic hypersurface. In case of the approximation of T this mixed approximation is the only possibility to cover an arbitrary compact subset of $A(x_0)$. The mixed approximation is particularly appropriate to approximate the basin of attraction step by step.

5.1 Approximation via a Single Operator

In this section we consider the approximation of a single operator. In Section 5.1.1 we approximate the function V satisfying $V'(x) = -p(x)$ via the operator $Dq(x) = q'(x)$ of the orbital derivative, where we follow [24]. In Section 5.1.2 we approximate the function $W(x) = \frac{V(x)}{\mathfrak{n}(x)}$ via the operator $D_m W(x) = W'(x) + m(x)W(x)$.

5.1.1 Approximation via Orbital Derivatives

Approximating the function $V'(x) = -p(x)$, cf. Theorem 2.46, we obtain an error estimate for the values of v and show in Theorem 5.1 a converse theorem to Theorem 2.24. However, by this theorem we will need grid points near the equilibrium which may lead to difficulties in the numerical calculation.

Theorem 5.1. *Let x_0 be an equilibrium of $\dot{x} = f(x)$ where $f \in C^\sigma(\mathbb{R}^n, \mathbb{R}^n)$ such that the real parts of all eigenvalues of $Df(x_0)$ are negative. Moreover, assume that $\sup_{x \in A(x_0)} \|f(x)\| < \infty$ or, more generally, $\sup_{x \in \mathbb{R}^n} \|f(x)\| < \infty$ holds; this can be achieved using (2.2).*

We consider the radial basis function $\Psi(x) = \psi_{l,k}(c\|x\|)$ with $c > 0$, where $\psi_{l,k}$ denotes the Wendland function with $k \in \mathbb{N}$ and $l := \lfloor \frac{n}{2} \rfloor + k + 1$. Let $\mathbb{N} \ni \sigma \geq \sigma^* := \frac{n+1}{2} + k$. Let $\mathfrak{q}$ be a (local) Lyapunov function with Lyapunov basin $\tilde{K} := \tilde{K}_r^{\mathfrak{q}}(x_0)$. Let $K_0 \subset A(x_0)$ be a compact set with $\tilde{K} \subset \overset{\circ}{K}_0$.

Then there is an open set B with $\overline{B} \subset A(x_0)$ and an $h^* > 0$, such that for all reconstructions $v \in C^{2k-1}(\mathbb{R}^n, \mathbb{R})$ of the Lyapunov function V of Theorem 2.46 where $V'(x) = -p(x)$ with respect to the grid $X_N \subset \overline{B} \setminus \{x_0\}$ with fill distance $h \leq h^*$ there is an extension v^* as in Theorem 4.8 and a compact set $K \supset K_0$ such that

- *$(v^*)'(x) < 0$ holds for all $x \in K \setminus \{x_0\}$,*
- *$K = \{x \in B \mid v^*(x) \leq (R^*)^2\}$ for an $R^* \in \mathbb{R}^+$.*

In other words, v^* is a Lyapunov function with Lyapunov basin K.

PROOF: Let $V \in C^\sigma(A(x_0), \mathbb{R})$ be the function of Theorem 2.46 which satisfies $V'(x) = -p(x)$ for all $x \in A(x_0)$ and $V(x_0) = 0$. Set $R := \sqrt{\max_{x \in K_0} V(x)} > 0$ and

$$\begin{aligned} K_1 &= \{x \in A(x_0) \mid V(x) \leq R^2\}, \\ K_2 &= \{x \in A(x_0) \mid V(x) \leq R^2 + 2\}, \\ B &= \{x \in A(x_0) \mid V(x) < R^2 + 3\}. \end{aligned}$$

Then obviously $K_0 \subset K_1 \subset K_2 \subset B \subset \overline{B} \subset A(x_0)$ and B is open, cf. Theorem 2.46; note that $\sup_{x \in A(x_0)} \|f(x)\| < \infty$. All these sets are positively invariant. Let $\tilde{B}$ be an open set with $\overline{B} \subset \tilde{B} \subset \overline{\tilde{B}} \subset A(x_0)$, e.g. $\tilde{B} = \{x \in A(x_0) \mid V(x) < R^2 + 4\}$. Let $\chi \in C_0^\infty(\mathbb{R}^n, [0,1])$ be a function with $\chi(x) = 1$ for $x \in \overline{B}$ and $\chi(x) = 0$ for $\mathbb{R}^n \setminus \tilde{B}$. Thus, $\chi \in C_0^\infty(\mathbb{R}^n) \subset \mathcal{F}$. Set $\tilde{a} := \|\chi\|_{\mathcal{F}}$ and $V_0 = V \cdot \chi$; then $V_0 \in C_0^\sigma(\mathbb{R}^n, \mathbb{R})$ and $V_0(x) = V(x)$ holds for all $x \in \overline{B}$. Lemma 3.13 implies $V_0 \in \mathcal{F}$. Choose $r_0' > 0$ so small that $\overline{B_{r_0'}(x_0)} = \{x \in \mathbb{R}^n \mid \|x - x_0\| \leq r_0'\} \subset \tilde{K}$ and

$$2r_0' \max_{\tilde{r} \in [0, r_0']} \left| \frac{d}{dr} \psi(\tilde{r}) \right| \leq \frac{1}{4\|V_0\|_{\mathcal{F}}^2} \tag{5.1}$$

hold where $\psi(r) := \psi_{l,k}(cr)$. This is possible since $\frac{d}{dr}\psi(r) = O(r)$ for $r \to 0$, cf. Proposition 3.11. Choose $r_0 > 0$ such that

$$\Omega := \{x \in A(x_0) \mid V_0(x) = r_0^2\} \subset \overline{B_{r_0'}(x_0)}$$

holds. Ω is a non-characteristic hypersurface by Lemma 2.37 and, hence, by Theorem 2.38 there exists a function $\theta \in C^\sigma(A(x_0) \setminus \{x_0\}, \mathbb{R})$ defined implicitly by $S_{\theta(x)} x \in \Omega$.

Set $\theta_0 := \max_{x \in \overline{B}} \theta(x) > 0$. Define $\min_{x \in \overline{B} \setminus \tilde{B}_r^q(x_0)} p(x) =: M_0 > 0$. Define $h^* > 0$ such that $h^* < \left(\frac{1}{C^*} \min\left(\frac{1}{2\theta_0}, M_0 \right) \right)^{\frac{1}{\kappa}}$ holds with C^* and κ as in Theorem 4.2 where $K_0 = \overline{B}$.

Let $X_N \subset \overline{B} \setminus \{x_0\}$ be a grid with fill distance $h \leq h^*$. For the approximant v of $V_0 (= V$ in $\overline{B})$ we set $\tilde{b} := -v(x_0)$. For the function $\tilde{v} := v + \tilde{b} \cdot \chi$ we have $\tilde{v}(x_0) = v(x_0) + \tilde{b} = 0$. For $x^* \in \overline{B_{r_0'}(x_0)}$ we have with $\delta_{x^*}, \delta_{x_0} \in \mathcal{F}^*$, cf. Lemma 3.22,

$$\begin{aligned} |V_0(x^*) - \tilde{v}(x^*)| &= |(\delta_{x^*} - \delta_{x_0})(V_0 - v - \tilde{b}\chi)| \\ &= |(\delta_{x^*} - \delta_{x_0})(V_0 - v)| \\ &\leq \|\delta_{x^*} - \delta_{x_0}\|_{\mathcal{F}^*} \cdot \|V_0 - v\|_{\mathcal{F}} \\ &\leq \|\delta_{x^*} - \delta_{x_0}\|_{\mathcal{F}^*} \cdot \|V_0\|_{\mathcal{F}} \text{ by Proposition 3.34.} \end{aligned} \tag{5.2}$$

Moreover, the Taylor expansion yields the existence of an $\tilde{r} \in [0, \rho]$, where $\rho := \|x^* - x_0\| \leq r_0'$ such that

$$\begin{aligned}\|\delta_{x^*} - \delta_{x_0}\|^2_{\mathcal{F}^*} &= (\delta_{x^*} - \delta_{x_0})^x (\delta_{x^*} - \delta_{x_0})^y \Psi(x-y) \\ &= (\delta_{x^*} - \delta_{x_0})^x \left[\psi(\|x - x^*\|) - \psi(\|x - x_0\|)\right] \\ &= 2\left[\psi(0) - \psi(\|x^* - x_0\|)\right] \\ &= -2\frac{d}{dr}\psi(\tilde{r})\rho \\ &\leq \frac{1}{4\|V_0\|^2_{\mathcal{F}}}, \text{ cf. (5.1).}\end{aligned}$$

Hence, we have with (5.2)

$$|V_0(x^*) - \tilde{v}(x^*)| \leq \frac{1}{2} \qquad \text{for all } x^* \in \overline{B_{r_0'}(x_0)}.$$

For $x \in \Omega$, i.e. $V(x) = V_0(x) = r_0^2$, we have $x \in \overline{B_{r_0'}(x_0)}$ and hence $\tilde{v}(x) \leq V_0(x) + \frac{1}{2} = r_0^2 + \frac{1}{2}$ and $\tilde{v}(x) \geq V_0(x) - \frac{1}{2} = r_0^2 - \frac{1}{2}$. For $v(x) = \tilde{v}(x) - \tilde{b}\chi(x)$ we thus have

$$v(x) \in \left[r_0^2 - \tilde{b} - \frac{1}{2}, r_0^2 - \tilde{b} + \frac{1}{2}\right] \qquad \text{for all } x \in \Omega. \tag{5.3}$$

For the orbital derivatives we have, using Theorem 4.2

$$|v'(x) - V_0'(x)| = |v'(x) + p(x)| \leq C^* h^\kappa =: \iota \text{ for all } x \in \overline{B}. \tag{5.4}$$

Since $C^* h^\kappa = \iota < M_0$ by definition of h^*, we have $v'(x) < -p(x) + M_0 \leq 0$ for all $x \in \overline{B} \setminus \tilde{B}_r^{\mathfrak{q}}(x_0)$. Hence, we can apply Theorem 4.8 and obtain an extension v^* of v, such that $v^*(x) = av(x) + b$ holds for all $x \in \overline{B} \setminus \tilde{B}_r^{\mathfrak{q}}(x_0)$ and $(v^*)'(x) < 0$ holds for all $x \in \overline{B} \setminus \{x_0\}$. Now set

$$\begin{aligned}K &= \{x \in B \mid v^*(x) \leq a(R^2 + 1 - \tilde{b}) + b =: (R^*)^2\} \\ &= \{x \in B \setminus \tilde{K} \mid v(x) \leq R^2 + 1 - \tilde{b}\} \cup \tilde{K}.\end{aligned}$$

The equation follows from the fact that $\tilde{K}$ is a subset of both sets, for the proof see below.

We will show that $K_1 \subset K \subset K_2$ holds. Then $K_0 \subset K$, K is a compact set and $(v^*)'(x) < 0$ holds for all $x \in K \setminus \{x_0\}$.

We show $K_1 \subset K$. For $x \in K_1 \setminus \tilde{B}_r^{\mathfrak{q}}(x_0)$ we have in particular $\theta_0 \geq \theta(x) \geq 0$ and with $\iota < \frac{1}{2\theta_0}$ we obtain

$$\begin{aligned}
v(x) &= v(S_{\theta(x)}x) - \int_0^{\theta(x)} v'(S_\tau x)\, d\tau \\
&\le r_0^2 - \tilde{b} + \frac{1}{2} + \int_0^{\theta(x)} (-V_0'(S_\tau x) + \iota)\, d\tau \text{ by (5.3) and (5.4)} \\
&\le \underbrace{V_0(S_{\theta(x)}x) - \int_0^{\theta(x)} V_0'(S_\tau x)\, d\tau}_{=V_0(x)} + \frac{1}{2} + \theta(x)\iota - \tilde{b} \\
&\le V_0(x) + \frac{1}{2} + \theta_0\iota - \tilde{b} \\
&\le R^2 + 1 - \tilde{b}, \text{ i.e. } x \in K.
\end{aligned}$$

Since in particular for $x \in \partial\tilde{K}$ the inequality $v(x) \le R^2+1-\tilde{b}$ and thus $v^*(x) \le a(R^2+1-\tilde{b})+b$ holds true and, moreover, v^* decreases along solutions, $\tilde{K} \subset K$ follows. In particular we have $a(R^2+1-\tilde{b})+b > 0$ since $v^*(x_0) = \mathfrak{q}(x_0) \ge 0$. Altogether, we have $K_1 \subset K$.

For the inclusion $K \subset K_2$ we show that for $x \in B \setminus K_2$ the inequality $v(x) > R^2 + 1 - \tilde{b}$ and thus $v^*(x) > a(R^2 + 1 - \tilde{b}) + b$ holds true. If $x \in B \setminus K_2 \subset A(x_0)$, then we have $\theta(x) \le \theta_0$ and

$$\begin{aligned}
v(x) &= v(S_{\theta(x)}x) - \int_0^{\theta(x)} v'(S_\tau x)\, d\tau \\
&\ge r_0^2 - \tilde{b} - \frac{1}{2} + \int_0^{\theta(x)} (-V_0'(S_\tau x) - \iota)\, d\tau \text{ by (5.3) and (5.4)} \\
&\ge \underbrace{V_0(S_{\theta(x)}x) - \int_0^{\theta(x)} V_0'(S_\tau x)\, d\tau}_{=V_0(x)} - \frac{1}{2} - \theta(x)\iota - \tilde{b} \\
&\ge V_0(x) - \frac{1}{2} - \theta_0\iota - \tilde{b} \\
&> R^2 + 2 - 1 - \tilde{b}, \text{ i.e. } x \notin K
\end{aligned}$$

since $\iota < \frac{1}{2\theta_0}$. This proves the theorem. □

5.1.2 Taylor Polynomial

In the following theorem we consider the function V where $V'(x) = -\|x - x_0\|^2$. We do not approximate V by this equation for the orbital derivative, but we approximate the function $W(x) = \frac{V(x)}{\mathfrak{n}(x)}$ as in Section 4.2.3 which satisfies $D_m W(x) = -\frac{\|x-x_0\|^2}{\mathfrak{n}(x)}$. The proof is similar to the one of Theorem 5.1.

Theorem 5.2. *Let x_0 be an equilibrium of $\dot{x} = f(x)$, where $f \in C^\sigma(\mathbb{R}^n, \mathbb{R}^n)$ such that the real parts of all eigenvalues of $Df(x_0)$ are negative. Moreover,* *assume that* $\sup_{x\in A(x_0)} \|f(x)\| < \infty$ *or, more generally,* $\sup_{x\in\mathbb{R}^n} \|f(x)\| < \infty$ *holds; this can be achieved using (2.2).*

We consider the radial basis function $\Psi(x) = \psi_{l,k}(c\|x\|)$ *with* $c > 0$, *where* $\psi_{l,k}$ *denotes the Wendland function with* $k \in \mathbb{N}$ *and* $l := \lfloor \frac{n}{2} \rfloor + k + 1$. *Let* V *be the Lyapunov function of Theorem 2.46 with* $V'(x) = -\|x - x_0\|^2$ *and* $V(x_0) = 0$, *and* $\mathfrak{n}(x) = \sum_{2\le|\alpha|\le P} c_\alpha (x-x_0)^\alpha + M\|x-x_0\|^{2H}$ *as in Definition 2.56, and let* $W(x) = \frac{V(x)}{\mathfrak{n}(x)} \in C^{P-2}(A(x_0), \mathbb{R})$ *with* $W(x_0) = 1$, *cf. Proposition 2.58. Let* $\sigma \ge P \ge 2+\sigma^*$, *where* $\sigma^* := \frac{n+1}{2} + k$. *Let* $K_0 \subset A(x_0)$ *be a compact set with* $x_0 \in \overset{\circ}{K}_0$.

Then there is an open set B *with* $\overline{B} \subset A(x_0)$, *such that for all reconstructions* $w \in C^{2k-1}(\mathbb{R}^n, \mathbb{R})$ *of* W *with respect to a grid* $X_N \subset \overline{B} \setminus \{x_0\}$ *which is dense enough in the sense of Theorem 4.10, there is a compact set* $K \supset K_0$ *such that with* $v_W(x) := \mathfrak{n}(x)w(x)$

- $v_W'(x) < 0$ *holds for all* $x \in K \setminus \{x_0\}$,
- $K = \{x \in B \mid v_W(x) \le (R^*)^2\}$ *for an* $R^* \in \mathbb{R}^+$.

In other words, v_W *is a Lyapunov function with Lyapunov basin* K.

Proof: Let $V \in C^\sigma(A(x_0), \mathbb{R})$ be the function of Theorem 2.46 which satisfies $V'(x) = -\|x - x_0\|^2$ for all $x \in A(x_0)$ and $V(x_0) = 0$. Then $W \in C^{P-2}(A(x_0), \mathbb{R})$ satisfies $W'(x) + \frac{\mathfrak{n}'(x)}{\mathfrak{n}(x)} W(x) = -\frac{\|x-x_0\|^2}{\mathfrak{n}(x)}$ for all $x \in A(x_0) \setminus \{x_0\}$, cf. (4.21). Note that $P - 2 \ge \sigma^*$. Set $R := \sqrt{\max_{x\in K_0} V(x)} > 0$ and

$$\begin{aligned} K_1 &= \{x \in A(x_0) \mid V(x) \le R^2\},\\ K_2 &= \{x \in A(x_0) \mid V(x) \le R^2 + 2\},\\ B &= \{x \in A(x_0) \mid V(x) < R^2 + 3\}. \end{aligned}$$

Then obviously $K_0 \subset K_1 \subset K_2 \subset B \subset \overline{B} \subset A(x_0)$ and B is open, cf. Theorem 2.46; note that $\sup_{x\in A(x_0)} \|f(x)\| < \infty$. All these sets are positively invariant. Let $\tilde{B}$ be an open set with $\overline{B} \subset \tilde{B} \subset \overline{\tilde{B}} \subset A(x_0)$, e.g. $\tilde{B} = \{x \in A(x_0) \mid V(x) < R^2 + 4\}$. Let $\chi \in C_0^\infty(\mathbb{R}^n, [0,1])$ be a function with $\chi(x) = 1$ for $x \in \overline{B}$ and $\chi(x) = 0$ for $\mathbb{R}^n \setminus \tilde{B}$. Thus, $\chi \in C_0^\infty(\mathbb{R}^n) \subset \mathcal{F}$. Set $\tilde{a} := \|\chi\|_\mathcal{F}$ and $W_0 = W \cdot \chi$; then $W_0 \in C_0^{P-2}(\mathbb{R}^n, \mathbb{R})$ and $W_0(x) = W(x)$ holds for all $x \in \overline{B}$. Lemma 3.13 implies $W_0 \in \mathcal{F}$. Choose $r_0' > 0$ so small that $\overline{B_{r_0'}(x_0)} \subset K_0$,

$$r_0' \le \frac{1}{\left(4C\sqrt{\Psi(0)}\|W_0\|_\mathcal{F}\right)^{\frac{1}{2}}} \tag{5.5}$$

$$\text{and } 2(r_0')^5 \max_{\tilde{r}\in[0,r_0']} \left|\frac{d}{dr}\psi(\tilde{r})\right| \le \frac{1}{(4C\|W_0\|_\mathcal{F})^2} \tag{5.6}$$

hold, where $\mathfrak{n}(x) \le C\|x - x_0\|^2$ for all $x \in \overline{B}$, cf. Proposition 2.58, 2., and $\psi(r) := \psi_{l,k}(cr)$. This is possible since $\frac{d}{dr}\psi(r) = O(r)$ for $r \to 0$, cf. Proposition 3.11. Choose $r_0 > 0$ such that

$$\Omega := \{x \in A(x_0) \mid V(x) = r_0^2\} \subset \overline{B_{r_0'}(x_0)}$$

holds. Ω is a non-characteristic hypersurface by Lemma 2.37 and hence, by Theorem 2.38 there exists a function $\theta \in C^\sigma(A(x_0)\setminus\{x_0\}, \mathbb{R})$ defined implicitly by $S_{\theta(x)}x \in \Omega$. Set $\theta_0 := \max_{x\in\overline{B}} \theta(x) > 0$. Let

$$\tilde{c} < \min\left(\frac{1}{2\theta_0 C M_0}, \frac{1}{C}\right), \tag{5.7}$$

where $M_0 := \max_{x\in\overline{B}} \|x - x_0\|^2$ and choose a grid $X_N \subset \overline{B} \setminus \{x_0\}$ according to Theorem 4.10.

For the approximant w of $W_0(= W \text{ in } \overline{B})$ we set $\tilde{b} := 1 - w(x_0)$. With $\delta_{x_0} \in \mathcal{F}^*$, cf. Lemma 3.25, and $W_0(x_0) = 1$, cf. Proposition 2.58, 3., we have

$$\begin{aligned}
|\tilde{b}| &= |\delta_{x_0}(W_0 - w)| \\
&\leq \|\delta_{x_0}\|_{\mathcal{F}^*} \cdot \|W_0 - w\|_{\mathcal{F}} \\
&\leq \sqrt{\Psi(0)} \cdot \|W_0\|_{\mathcal{F}}
\end{aligned} \tag{5.8}$$

by Proposition 3.37 since $\|\delta_{x_0}\|^2_{\mathcal{F}^*} = \delta^x_{x_0}\delta^y_{x_0}\Psi(x-y) = \Psi(0)$. For the function $\tilde{w} := w + \tilde{b}\cdot\chi$ we have $\tilde{w}(x_0) = w(x_0) + \tilde{b} = 1$. For $x^* \in \overline{B_{r_0'}(x_0)}$ we have thus

$$\begin{aligned}
|W_0(x^*) - \tilde{w}(x^*)| &= |(\delta_{x^*} - \delta_{x_0})(W_0 - w - \tilde{b}\cdot\chi)| \\
&= |(\delta_{x^*} - \delta_{x_0})(W_0 - w)| \\
&\leq \|\delta_{x^*} - \delta_{x_0}\|_{\mathcal{F}^*} \cdot \|W_0 - w\|_{\mathcal{F}} \\
&\leq \|\delta_{x^*} - \delta_{x_0}\|_{\mathcal{F}^*} \cdot \|W_0\|_{\mathcal{F}} \text{ by Proposition 3.37.}
\end{aligned}$$

Moreover, the Taylor expansion yields the existence of an $\tilde{r} \in [0, \rho]$ where $\rho := \|x^* - x_0\| \leq r_0'$ such that

$$\begin{aligned}
\|\delta_{x^*} - \delta_{x_0}\|^2_{\mathcal{F}^*} &= (\delta_{x^*} - \delta_{x_0})^x(\delta_{x^*} - \delta_{x_0})^y\Psi(x-y) \\
&= (\delta_{x^*} - \delta_{x_0})^x\left[\psi(\|x - x^*\|) - \psi(\|x - x_0\|)\right] \\
&= 2\left[\psi(0) - \psi(\|x^* - x_0\|)\right] \\
&= -2\psi'(\tilde{r})\rho \\
&\leq \frac{1}{(4C\cdot(r_0')^2\|W_0\|_{\mathcal{F}})^2}, \text{ cf. (5.6).}
\end{aligned}$$

Hence, for all $x^* \in \overline{B_{r_0'}(x_0)}$ we have

$$|W_0(x^*) - \tilde{w}(x^*)| \leq \frac{1}{4C\cdot(r_0')^2}. \tag{5.9}$$

For $v_W(x) = \mathfrak{n}(x)w(x) = \mathfrak{n}(x)[\tilde{w}(x) - \tilde{b}\chi(x)]$ we have for all $x^* \in \overline{B_{r_0'}(x_0)}$

$$\begin{aligned}|V(x^*) - v_W(x^*)| &= \mathfrak{n}(x^*)[W_0(x^*) - \tilde{w}(x^*) + \tilde{b}\chi(x^*)]\\ &\leq \underbrace{\max_{x \in B_{r_0'}(x_0)} \mathfrak{n}(x)}_{\leq C \cdot (r_0')^2} \left[|W_0(x^*) - \tilde{w}(x^*)| + |\tilde{b}|\right]\\ &\leq \frac{1}{4} + \frac{1}{4} \qquad \text{by (5.5), (5.8) and (5.9).}\end{aligned}$$

Thus,

$$v_W(x) \in \left[r_0^2 - \frac{1}{2}, r_0^2 + \frac{1}{2}\right] \qquad \text{for all } x \in \Omega. \tag{5.10}$$

For the orbital derivatives we have, using Theorem 4.10, (4.18) and (4.19)

$$v_W'(x) < 0 \qquad \text{for all } x \in \overline{B} \setminus \{x_0\} \text{ and} \tag{5.11}$$

$$|v_W'(x) + \|x - x_0\|^2| \leq \tilde{c}C\|x - x_0\|^2 \ \leq\ \tilde{c}CM_0 \ \leq\ \frac{1}{2\theta_0} \tag{5.12}$$

by (5.7) for all $x \in \overline{B}$.

Now set

$$K = \{x \in B \mid v_W(x) \leq R^2 + 1 =: (R^*)^2\}.$$

We will show that $K_1 \subset K \subset K_2$ holds. Then $K_0 \subset K$, K is a compact set and $v_W'(x) < 0$ holds for all $x \in K \setminus \{x_0\}$, cf. (5.11).

We show $K_1 \subset K$. Let $x \in K_1$. We distinguish between the cases $\theta(x) < 0$ and $\theta(x) \geq 0$. If $\theta(x) < 0$, then

$$\begin{aligned}v_W(x) &= v_W(S_{\theta(x)}x) - \int_0^{\theta(x)} v_W'(S_\tau x)\, d\tau\\ &\leq v_W(S_{\theta(x)}x)\\ &\leq r_0^2 + \frac{1}{2} \text{ by (5.10)}\\ &\leq R^2 + 1,\end{aligned}$$

since $R^2 = \max_{x \in K_0} V(x) \geq \max_{x \in \Omega} V(x) = r_0^2$.

Now assume $\theta_0 \geq \theta(x) \geq 0$. We have

$$\begin{aligned}v_W(x) &= v_W(S_{\theta(x)}x) - \int_0^{\theta(x)} v_W'(S_\tau x)\, d\tau\\ &\leq r_0^2 + \frac{1}{2} + \int_0^{\theta(x)} \left(\|S_\tau x - x_0\|^2 + \frac{1}{2\theta_0}\right) d\tau \text{ by (5.10) and (5.12)}\\ &\leq \underbrace{V(S_{\theta(x)}x) - \int_0^{\theta(x)} V'(S_\tau x)\, d\tau}_{=V(x)} + \frac{1}{2} + \frac{\theta(x)}{2\theta_0}\\ &\leq V(x) + 1\\ &\leq R^2 + 1,\end{aligned}$$

i.e. $x \in K$. Hence, $K_1 \subset K$.

For the inclusion $K \subset K_2$ we show that for $x \in B \setminus K_2$ the inequality $v_W(x) > R^2+1$ holds true. If $x \in B \setminus K_2 \subset A(x_0)$, then we have $0 \leq \theta(x) \leq \theta_0$ and

$$
\begin{aligned}
v_W(x) &= v_W(S_{\theta(x)}x) - \int_0^{\theta(x)} v_W'(S_\tau x)\,d\tau \\
&\geq r_0^2 - \frac{1}{2} + \int_0^{\theta(x)} \left(\|S_\tau x - x_0\|^2 - \frac{1}{2\theta_0}\right) d\tau \text{ by (5.10) and (5.12)} \\
&\geq \underbrace{V(S_{\theta(x)}x) - \int_0^{\theta(x)} V'(S_\tau x)\,d\tau}_{=V(x)} - \frac{1}{2} - \frac{\theta(x)}{2\theta_0} \\
&\geq V(x) - \frac{1}{2} - \frac{1}{2} \\
&> R^2 + 2 - 1,
\end{aligned}
$$

i.e. $x \notin K$. This proves the theorem. □

5.2 Mixed Approximation

For T (and also for V) one can use a mixed approximation. Here, additionally to the orbital derivative Q', the values of Q are given on an $(n-1)$-dimensional manifold, a *non-characteristic hypersurface*. Such a non-characteristic hypersurface can be given by the level set of a (local) Lyapunov function within its Lyapunov basin. With this method one can also cover each compact subset of the basin of attraction by a Lyapunov basin when approximating the function T or V. In the case of T, where $T'(x) = -1$, the level sets of the function T and thus also of t up to a certain error have a special meaning: a solution needs the time $T_2 - T_1$ from the level set $T = T_2$ to the level set $T = T_1$.

Moreover, one can exhaust the basin of attraction by compact sets: starting with a local Lyapunov function and a corresponding local Lyapunov basin K_0, one obtains a larger Lyapunov basin K_1 through mixed approximation using the boundary ∂K_0 as a non-characteristic hypersurface. The boundary ∂K_1 is again a non-characteristic hypersurface and hence one obtains a sequence of compact sets $K_0 \subset K_1 \subset \ldots$ which exhaust $A(x_0)$. In Figure 5.4 we show the first step of this method with the function V^* for example (2.11): starting with a local Lyapunov basin K_0 (magenta), we obtain a larger Lyapunov basin K_1 using mixed approximation. In [23] the same example with a different grid X_N is considered, and one more step is calculated.

In this section we approximate the function $Q = T$ or $Q = V^*$ via its orbital derivatives and its function values. The orbital derivatives are given on

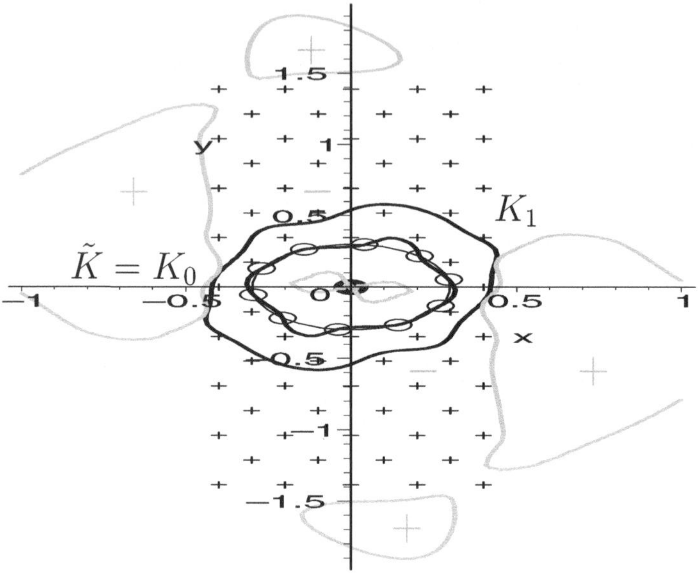

Fig. 5.4. Mixed approximation of V^* where $(V^*)'(x,y) = -(x^2+y^2)$ with given value $V^*(x,y) = 1$ on the boundary of the local Lyapunov basin $\tilde{K} = K_0$ (thin black). We used a grid of $M = 10$ points (black circles) on $\partial\tilde{K}$ for the values of the approximation v and a second grid of $N = 70$ points (black +) with $\alpha = 0.2$ for the orbital derivative v'. The sign of $v'(x,y)$ (grey), and the level sets $v(x,y) = 1$ and $v(x,y) = 1.1$ (black) are shown. The level set $v(x,y) = 1.1$ is the boundary of K_1 (black) which is a subset of the basin of attraction for (2.11).

a grid X_N, whereas the function values are given on a different grid $X_M^0 \subset \Omega$ where Ω is a non-characteristic hypersurface. In most cases, Ω is given by a level set of a Lyapunov function $\mathfrak{q}$, e.g. a local Lyapunov function. This mixed interpolation problem was discussed in Section 3.1.3, cf. Definition 3.7. Any compact subset $K_0 \subset A(x_0)$ can be covered by a Lyapunov basin K obtained by a mixed approximation via radial basis functions as we prove in Section 5.2.1.

Moreover, we can approach the basin of attraction stepwise by a sequence of Lyapunov functions q_i, $i = 1, 2, \ldots$ with Lyapunov basins $K_i \supset K_{i-1}$. The advantage of this approach is that one can use a grid X_N outside K_{i-1} in each step, cf. Section 5.2.2.

5.2.1 Approximation via Orbital Derivatives and Function Values

We approximate the Lyapunov function T satisfying $T' = -\bar{c}$. Note that we fix the values of T on the boundary of a Lyapunov basin, i.e. on a non-characteristic hypersurface.

Theorem 5.3. *Consider the function $\Psi(x) = \psi_{l,k}(c\|x\|)$ with $c > 0$, where $\psi_{l,k}$ denotes the Wendland function with $k \in \mathbb{N}$ and $l := \lfloor \frac{n}{2} \rfloor + k + 1$. Let $f \in C^\sigma(\mathbb{R}^n, \mathbb{R}^n)$, where $\mathbb{N} \ni \sigma \geq \sigma^* := \frac{n+1}{2} + k$. Moreover, assume that $\sup_{x \in A(x_0)} \|f(x)\| < \infty$ or, more generally, $\sup_{x \in \mathbb{R}^n} \|f(x)\| < \infty$ holds. Let $\mathfrak{q}$ be a Lyapunov function with Lyapunov basin $\tilde{K} := \tilde{K}_r^{\mathfrak{q}}(x_0)$ according to Definition 2.23. Define $\Omega := \partial\tilde{K}$. Let K_0 be a compact set with $\tilde{K} \subset \overset{\circ}{K}_0 \subset K_0 \subset A(x_0)$ and let $H \in C^\sigma(\Omega, \mathbb{R}_0^+)$.*

Then there is an open set B with $\overline{B} \subset A(x_0)$ and constants $h_0^*, h^* > 0$ such that for every reconstruction t of T in the sense of Definition 3.7, where T is defined in Theorem 2.38 with $T(x) = H(x)$ for $x \in \Omega$, using grids $X_N \subset \overline{B} \setminus \tilde{B}_r^{\mathfrak{q}}(x_0)$ with fill distance $h \leq h^*$ and $X_M^0 \subset \Omega$ with fill distance $h_0 \leq h_0^*$, there is an extension $t^* \in C^{2k-1}(\mathbb{R}^n, \mathbb{R})$ of t as in the Extension Theorem 4.8, such that:

There is a compact set $B \supset K \supset K_0$ with

1. *$K = \{x \in B \mid t^*(x) \leq (R^*)^2\}$ for an $R^* \in \mathbb{R}^+$,*
2. *$(t^*)'(x) < 0$ for all $x \in K \setminus \{x_0\}$.*

In other words, t^* is a Lyapunov function with Lyapunov basin K.

PROOF: We assume without loss of generality that $\mathfrak{q}(x_0) = 0$. Ω is a non-characteristic hypersurface and we define the function θ for all $x \in A(x_0) \setminus \{x_0\}$ implicitly by

$$S_{\theta(x)} x \in \Omega, \text{ i.e. } \mathfrak{q}(S_{\theta(x)} x) = r^2,$$

cf. Theorem 2.38, and set

$$\theta_0 := \max_{x \in K_0 \setminus \tilde{B}_r^{\mathfrak{q}}(x_0)} \theta(x) \geq 0,$$

$$K_1 := \left\{ x \in A(x_0) \setminus \tilde{B}_r^{\mathfrak{q}}(x_0) \mid \theta(x) \leq \theta_0 \right\} \cup \tilde{B}_r^{\mathfrak{q}}(x_0).$$

Then obviously $K_0 \subset K_1$, and K_1 is positively invariant.

Define $T \in C^\sigma(A(x_0) \setminus \{x_0\}, \mathbb{R})$ as in Theorem 2.38, i.e. $T'(x) = -\bar{c}$ for $x \in A(x_0) \setminus \{x_0\}$ and $T(x) = H(x)$ for $x \in \Omega$. We set $c_M := \max_{x \in \Omega} H(x) = \max_{x \in \Omega} T(x)$ and $c_m := \min_{x \in \Omega} H(x) = \min_{x \in \Omega} T(x)$. With $\theta^* := \frac{c_M - c_m + \frac{3}{2}\bar{c}\,\theta_0 + 2}{\frac{1}{2}\bar{c}} > \theta_0$ we define the following sets

$$K_2 := \left\{ x \in A(x_0) \setminus \tilde{B}_r^{\mathfrak{q}}(x_0) \mid \theta(x) \leq \theta^* \right\} \cup \tilde{B}_r^{\mathfrak{q}}(x_0),$$

$$B := \left\{ x \in A(x_0) \setminus \tilde{B}_r^{\mathfrak{q}}(x_0) \mid \theta(x) < \theta^* + 1 \right\} \cup \tilde{B}_r^{\mathfrak{q}}(x_0).$$

Then obviously $K_1 \subset K_2 \subset B$, both K_2 and B are positively invariant and B is open, cf. Proposition 2.44; note that $\sup_{x \in A(x_0)} \|f(x)\| < \infty$.

We modify T in $\tilde{B}_r^{\mathfrak{q}}(x_0)$ and outside $\overline{B}$ such that $T \in C_0^\sigma(\mathbb{R}^n, \mathbb{R}) \subset \mathcal{F}$ and T remains unchanged in $\overline{B} \setminus \tilde{B}_r^{\mathfrak{q}}(x_0)$, cf. the proof of Theorem 4.1.

We apply Theorem 3.41 to $K = \overline{B} \setminus \tilde{B}_r^{\mathfrak{q}}(x_0)$ and to grids $X_N \subset \overline{B} \setminus \tilde{B}_r^{\mathfrak{q}}(x_0)$ with fill distance $h \leq \left(\frac{\bar{c}}{2C^*}\right)^{\frac{1}{\kappa}} =: h^*$ and $X_M^0 \subset \Omega$ with fill distance $h_0 \leq \frac{1}{C_0^*} =: h_0^*$, where C^* and C_0^* are as in Theorem 3.41. We obtain an approximation $t \in C^{2k-1}(\mathbb{R}^n, \mathbb{R})$ for which the following inequality holds for all $x \in \overline{B} \setminus \tilde{B}_r^{\mathfrak{q}}(x_0)$, cf. (3.49),

$$-\frac{3}{2}\bar{c} \leq t'(x) \leq -\frac{1}{2}\bar{c} < 0. \tag{5.13}$$

The Extension Theorem 4.8, applied to $K = \overline{B}$, $\mathfrak{q}$ and $q = t$ implies

$$(t^*)'(x) < 0 \qquad \text{for all } x \in \overline{B} \setminus \{x_0\}. \tag{5.14}$$

Note that for the function t^* we have $t^*(x) = at(x) + b$ for all $x \notin \tilde{B}_r^{\mathfrak{q}}(x_0)$. We set $R := \sqrt{c_M + 1 + \frac{3}{2}\bar{c}\,\theta_0}$ and define

$$\begin{aligned} K &:= \{x \in B \mid t^*(x) \leq aR^2 + b =: (R^*)^2\} \\ &= \{x \in B \setminus \tilde{B}_r^{\mathfrak{q}}(x_0) \mid t(x) \leq R^2\} \cup \tilde{B}_r^{\mathfrak{q}}(x_0). \end{aligned}$$

The equation follows from the fact that $\tilde{B}_r^{\mathfrak{q}}(x_0)$ is a subset of both sets, for the proof see below. Note that by (3.50) of Theorem 3.41 we have the following result for all $x \in B \setminus \{x_0\}$:

$$t(S_{\theta(x)}x) \in [c_m - 1, c_M + 1] \tag{5.15}$$

since $S_{\theta(x)}x \in \Omega$.

We will show that $K_1 \subset K \subset K_2$ holds. Note that $K_1 \subset K$ implies $K_0 \subset K$. $K \subset K_2 \subset B$, on the other hand, shows that K is a compact set; 2. then follows from (5.14).

We show $K_1 \subset K$: For $x \in K_1 \setminus \tilde{B}_r^{\mathfrak{q}}(x_0)$, we have with $0 \leq \theta(x) \leq \theta_0$, the positive invariance of K_1 and (5.15)

$$\begin{aligned} c_M + 1 &\geq t(S_{\theta(x)}x) \\ &= t(x) + \int_0^{\theta(x)} t'(S_\tau x)\, d\tau \\ &\geq t(x) - \frac{3}{2}\bar{c}\,\theta(x) \text{ by (5.13)} \\ t(x) &\leq c_M + 1 + \frac{3}{2}\bar{c}\,\theta_0 = R^2 \end{aligned}$$

and hence $x \in K$. For $x \in \tilde{B}_r^{\mathfrak{q}}(x_0)$, we have by the Extension Theorem 4.8 and (5.15) $t^*(x) \leq \max_{\xi \in \Omega} t^*(\xi) \leq a(c_M + 1) + b \leq (R^*)^2$. Thus, $x \in K$.

We show $K \subset K_2$: Assume in contradiction that there is an $x \in B \setminus K_2$ with $t(x) \leq R^2$ and $\theta(x) > \theta^*$ – note that $\tilde{B}_r^{\mathfrak{q}}(x_0) \subset K_2$ by construction. By (5.15) we have

$$
\begin{aligned}
c_m - 1 &\le t(S_{\theta(x)}x) \\
&= t(x) + \int_0^{\theta(x)} t'(S_\tau x)\, d\tau \\
&\le t(x) - \frac{1}{2}\bar{c}\,\theta(x) \text{ by (5.13)} \\
t(x) &> c_m - 1 + \frac{1}{2}\bar{c}\,\theta^* = R^2
\end{aligned}
$$

by definition of θ^*. This is a contradiction and thus $K \subset K_2$, which proves the theorem. □

The following corollary shows that the difference of the values of t corresponds to the time which a solution needs from one level set to another up to the error $\max_\xi |t'(\xi) + \bar{c}|$.

Corollary 5.4 *Let the assumptions of Theorem 5.3 hold. For $x \in \mathbb{R}^n$ and $\tilde{t} > 0$ let $S_\tau x \in K \setminus \tilde{B}_r^{\mathfrak{q}}(x_0)$ hold for all $\tau \in [0, \tilde{t}]$. Denote $\rho_1 := t(x)$ and $\rho_0 := t(S_{\tilde{t}}x)$. Moreover, let $\max_{\xi \in K \setminus \tilde{B}_r^{\mathfrak{q}}(x_0)} |t'(\xi) + \bar{c}| =: \iota < \bar{c}$ hold (by the assumptions of Theorem 5.3, in particular (5.13), $\iota = C^* h^\kappa \le \frac{\bar{c}}{2}$ is an upper bound).*

Then the time $\tilde{t}$ fulfills

$$
\frac{\rho_1 - \rho_0}{\bar{c} + \iota} \le \tilde{t} \le \frac{\rho_1 - \rho_0}{\bar{c} - \iota}.
$$

PROOF: We have $\rho_0 - \rho_1 = \int_0^{\tilde{t}} t'(S_\tau x)\, d\tau$. Since $|t'(S_\tau x) + \bar{c}| \le \iota$ holds for all $\tau \in [0, \tilde{t}]$, we have $(-\bar{c} - \iota)\,\tilde{t} \le \rho_0 - \rho_1 \le (-\bar{c} + \iota)\,\tilde{t}$, which proves the corollary. □

Now we consider the Lyapunov function V satisfying $V'(x) = -p(x)$. Fixing the values on a non-characteristic hypersurface Ω, we have to consider the function V^*, cf. Proposition 2.51, which satisfies $(V^*)'(x) = -p(x)$ for $x \in A(x_0) \setminus \{x_0\}$ and $V^*(x) = H(x)$ for $x \in \Omega$, where H is a given function.

Theorem 5.5. *Consider the function $\Psi(x) = \psi_{l,k}(c\|x\|)$ with $c > 0$, where $\psi_{l,k}$ denotes the Wendland function with $k \in \mathbb{N}$ and $l := \lfloor \frac{n}{2} \rfloor + k + 1$. Let $f \in C^\sigma(\mathbb{R}^n, \mathbb{R}^n)$, where $\mathbb{N} \ni \sigma \ge \sigma^* := \frac{n+1}{2} + k$. Moreover, assume that $\sup_{x \in A(x_0)} \|f(x)\| < \infty$ or, more generally, $\sup_{x \in \mathbb{R}^n} \|f(x)\| < \infty$ holds. Let $\mathfrak{q}$ be a Lyapunov function with Lyapunov basin $\tilde{K} := \tilde{K}_r^{\mathfrak{q}}(x_0)$ according to Definition 2.23. Define $\Omega := \partial\tilde{K}$. Let K_0 be a compact set with $\tilde{K} \subset \mathring{K}_0 \subset K_0 \subset A(x_0)$ and let $H \in C^\sigma(\Omega, \mathbb{R}_0^+)$.*

Then there is an open set B with $\overline{B} \subset A(x_0)$ and constants $h_0^*, h^* > 0$ such that for every reconstruction v of V^* in the sense of Definition 3.7, where V^* is defined in Proposition 2.51 with $V^*(x) = H(x)$ for $x \in \Omega$, using grids $X_N \subset \overline{B} \setminus \tilde{B}_r^{\mathfrak{q}}(x_0)$ with fill distance $h \le h^*$ and $X_M^0 \subset \Omega$ with fill distance $h_0 \le h_0^*$, there is an extension $v^* \in C^{2k-1}(\mathbb{R}^n, \mathbb{R})$ of v as in the Extension Theorem 4.8, such that:

There is a compact set $B \supset K \supset K_0$ with

1. $K = \{x \in B \mid v^*(x) \leq (R^*)^2$ *for an* $R^* \in \mathbb{R}^+$,
2. $(v^*)'(x) < 0$ *for all* $x \in K \setminus \{x_0\}$.

In other words, v^* is a Lyapunov function with Lyapunov basin K.

PROOF: We assume without loss of generality that $\mathfrak{q}(x_0) = 0$. Ω is a non-characteristic hypersurface and we define the function θ for all $x \in A(x_0) \setminus \{x_0\}$ implicitly by

$$S_{\theta(x)}x \in \Omega, \text{ i.e. } \mathfrak{q}(S_{\theta(x)}x) = r^2,$$

cf. Theorem 2.38, and set

$$\theta_0 := \max_{x \in K_0 \setminus \tilde{B}_r^{\mathfrak{q}}(x_0)} \theta(x) \geq 0,$$

$$K_1 := \left\{x \in A(x_0) \setminus \tilde{B}_r^{\mathfrak{q}}(x_0) \mid \theta(x) \leq \theta_0\right\} \cup \tilde{B}_r^{\mathfrak{q}}(x_0).$$

Then $K_0 \subset K_1$, and K_1 is positively invariant. We set $\epsilon := \frac{1}{2} \inf_{x \notin \tilde{B}_r^{\mathfrak{q}}(x_0)} p(x) > 0$, $p_M := \max_{x \in K_1} p(x)$, $c_M := \max_{x \in \Omega} H(x) = \max_{x \in \Omega} V^*(x)$ and $c_m := \min_{x \in \Omega} H(x) = \min_{x \in \Omega} V^*(x)$. With $\theta^* := \frac{c_M - c_m + (p_M + \epsilon)\theta_0 + 2}{\epsilon} > \theta_0$ we define the following sets

$$K_2 := \left\{x \in A(x_0) \setminus \tilde{B}_r^{\mathfrak{q}}(x_0) \mid \theta(x) \leq \theta^*\right\} \cup \tilde{B}_r^{\mathfrak{q}}(x_0),$$

$$B := \left\{x \in A(x_0) \setminus \tilde{B}_r^{\mathfrak{q}}(x_0) \mid \theta(x) < \theta^* + 1\right\} \cup \tilde{B}_r^{\mathfrak{q}}(x_0).$$

Then obviously $K_1 \subset K_2 \subset B$, both K_2 and B are positively invariant and B is open, cf. Proposition 2.51; note that $\sup_{x \in A(x_0)} \|f(x)\| < \infty$.

We modify V^* in $\tilde{B}_r^{\mathfrak{q}}(x_0)$ and outside $\overline{B}$ such that $V^* \in C_0^\sigma(\mathbb{R}^n, \mathbb{R}) \subset \mathcal{F}$ and V^* remains unchanged in $\overline{B} \setminus \tilde{B}_r^{\mathfrak{q}}(x_0)$, cf. the proof of Theorem 4.1.

We apply Theorem 3.41 to $K = \overline{B} \setminus \tilde{B}_r^{\mathfrak{q}}(x_0)$ and to grids $X_N \subset \overline{B} \setminus \tilde{B}_r^{\mathfrak{q}}(x_0)$ with fill distance $h \leq \left(\frac{\epsilon}{C^*}\right)^{\frac{1}{\kappa}} =: h^*$ and $X_M^0 \subset \Omega$ with fill distance $h_0 \leq \frac{1}{C_0^*} =: h_0^*$, where C^* and C_0^* are as in Theorem 3.41. We obtain a function $v \in C^{2k-1}(\mathbb{R}^n, \mathbb{R})$ for which the following inequality holds for all $x \in \overline{B} \setminus \tilde{B}_r^{\mathfrak{q}}(x_0)$, cf. (3.49),

$$-p(x) - \epsilon \leq v'(x) \leq -p(x) + \epsilon < 0. \tag{5.16}$$

The Extension Theorem 4.8, applied to $K = \overline{B}$, $\mathfrak{q}$ and $q = v$ implies

$$(v^*)'(x) < 0 \qquad \text{for all } x \in \overline{B} \setminus \{x_0\}. \tag{5.17}$$

Note that for the function v^* we have then $v^*(x) = av(x) + b$ for all $x \notin \tilde{B}_r^{\mathfrak{q}}(x_0)$. We set $R := \sqrt{c_M + 1 + (p_M + \epsilon)\theta_0}$ and define

$$\begin{aligned} K &:= \{x \in B \mid v^*(x) \leq aR^2 + b =: (R^*)^2\} \\ &= \{x \in B \setminus \tilde{B}_r^{\mathfrak{q}}(x_0) \mid v(x) \leq R^2\} \cup \tilde{B}_r^{\mathfrak{q}}(x_0). \end{aligned}$$

The equation follows from the fact that $\tilde{B}_r^{\mathfrak{q}}(x_0)$ is a subset of both sets, for the proof see below. Note that by (3.50) of Theorem 3.41 we have the following result for all $x \in B \setminus \{x_0\}$:

$$v(S_{\theta(x)}x) \in [c_m - 1, c_M + 1] \tag{5.18}$$

since $S_{\theta(x)}x \in \Omega$.

We will show that $K_1 \subset K \subset K_2$ holds. Note that $K_1 \subset \tilde{K}$ implies $K_0 \subset K$. $K \subset K_2 \subset B$, on the other hand, shows that K is a compact set; 2. then follows from (5.17).

We show $K_1 \subset K$: For $x \in K_1 \setminus \tilde{B}_r^{\mathfrak{q}}(x_0)$, we have with $0 \leq \theta(x) \leq \theta_0$, the positive invariance of K_1 and (5.18)

$$\begin{aligned}
c_M + 1 &\geq v(S_{\theta(x)}x) \\
&= v(x) + \int_0^{\theta(x)} \underbrace{v'(S_t x)}_{\geq -p(S_t x) - \epsilon} \, dt \quad \text{by (5.16)} \\
&\geq v(x) - (p_M + \epsilon)\theta(x) \\
v(x) &\leq c_M + 1 + (p_M + \epsilon)\theta_0 = R^2,
\end{aligned}$$

and hence $x \in K$. For $x \in \tilde{B}_r^{\mathfrak{q}}(x_0)$, we have by the Extension Theorem 4.8 and (5.18) $v^*(x) \leq \max_{\xi \in \Omega} v^*(\xi) \leq a(c_M + 1) + b \leq (R^*)^2$. Thus, $x \in K$.

We show $K \subset K_2$: Assume in contradiction that there is an $x \in B \setminus K_2$ with $v(x) \leq R^2$ and $\theta(x) > \theta^*$ – note that $\tilde{B}_r^{\mathfrak{q}}(x_0) \subset K_2$ by construction. By (5.18) we have

$$\begin{aligned}
c_m - 1 &\leq v(S_{\theta(x)}x) \\
&= v(x) + \int_0^{\theta(x)} \underbrace{v'(S_t x)}_{\leq -p(S_t x) + \epsilon} \, dt \quad \text{by (5.16)} \\
&\leq v(x) + (-2\epsilon + \epsilon)\theta(x) \\
v(x) &> c_m - 1 + \epsilon\theta^* = R^2
\end{aligned}$$

by definition of θ^*. This is a contradiction and thus $K \subset K_2$, which proves the theorem. □

5.2.2 Stepwise Exhaustion of the Basin of Attraction

Using Theorem 5.3 or Theorem 5.5 we can stepwise exhaust the basin of attraction, cf. also Section 6.3. We assume that $\sup_{x \in A(x_0)} \|f(x)\| < \infty$.

Calculate a local Lyapunov function $\mathfrak{q}$ and a corresponding local Lyapunov basin $\tilde{K} = \tilde{K}_r^{\mathfrak{q}}(x_0)$. Denote $q_0 := \mathfrak{q}$, $K_0 := \tilde{K}$ and $r_0 := r$, and set $B_0 = \mathbb{R}^n$. This is the departing point for a sequence of compact Lyapunov basins K_i, $i = 1, 2, \ldots$, with $K_{i+1} \supset K_i$ and $\bigcup_{i \in \mathbb{N}} K_i = A(x_0)$.

Now assume that a Lyapunov function q_i with Lyapunov basin $K_i = \tilde{K}^{q_i}_{r_i}(x_0)$ and neighborhood B_i is given. The only information we need of the Lyapunov function and the Lyapunov basin is the boundary $\partial K_i =: \Omega_{i+1}$. Hence, if q_i is the extension of a function $\tilde{q}_i$, it suffices to know the set ∂K_i either given by $\partial \tilde{K}^{\tilde{q}_i}_{\tilde{r}_i}(x_0) = \{x \in B_i \mid \tilde{q}_i(x) = \tilde{r}_i^2\}$ or by $\partial \tilde{K}^{q_i}_{r_i}(x_0) = \{x \in B_i \mid q_i(x) = r_i^2\}$.

Let $B_{i+1} \supset K_i$ be an open set which will be specified below. Choose grids $X_N \subset \overline{B_{i+1}} \setminus K_i$ – in practical applications we let the grid be slightly larger, also including points in K_i near ∂K_i – and $X^0_M \subset \Omega_{i+1} := \partial K_i$. Now approximate either $Q = T$ or $Q = V^*$ by a mixed approximation with respect to the grids X_N and X^0_M and the values $Q(\xi_j) = H(\xi_j) = 1$ for $\xi_j \in X^0_M$. Make the grids dense enough so that for the reconstruction q there is a set $S_{i+1} := \{x \in B_{i+1} \mid 1 - \epsilon_{i+1} \leq q(x) \leq \tilde{r}^2_{i+1}\}$ with $\epsilon_{i+1} > 0$ such that $q'(x) < 0$ holds for all $x \in S_{i+1}$ and $\partial K_i \subset \overset{\circ}{S}_{i+1}$. Then there is an extension q^* of q with $\mathfrak{q} = q_i$ such that $q_{i+1}(x) := q^*(x)$ is a Lyapunov function with Lyapunov basin $K_{i+1} := S_{i+1} \cup K_i = \{x \in B_{i+1} \mid q^*(x) \leq a\tilde{r}^2_{i+1} + b =: r^2_{i+1}\}$.

We show that with this method $\bigcup_{i \in \mathbb{N}} K_i = A(x_0)$ holds and we can thus stepwise exhaust the basin of attraction, if we choose B_{i+1} properly. To show this, we reprove the induction step from K_i to K_{i+1} again. Let K^*_i be a sequence of compact sets with $K^*_{i+1} \supset K^*_i$ and $\bigcup_{i \in \mathbb{N}} K^*_i = A(x_0)$, e.g. $K^*_i = S_{-i}\tilde{K}$ where S_{-i} denotes the flow and $\tilde{K}$ is the local Lyapunov basin defined above. By Proposition 2.44, the sets K^*_i are compact and since for all $z \in A(x_0)$ there is a finite time T^* with $S_{T^*} z \in \tilde{K}$, we obtain $\bigcup_{i \in \mathbb{N}} K^*_i = A(x_0)$.

Now we reprove the induction step from K_i to K_{i+1}. For given i choose $l_{i+1} > l_i$ so large that $K_i \subset \overset{\circ}{K}{}^*_{l_{i+1}} \subset K^*_{l_{i+1}} \subset A(x_0)$ holds. Such an l_{i+1} exists due to the compactness of K_i. Then Theorem 5.3 or 5.5 with $\mathfrak{q} = q_i$, $\tilde{K} = K_i$ and $K_0 = K^*_{l_{i+1}}$ implies that there is an open set $B =: B_{i+1}$ and a Lyapunov function q^* with Lyapunov basin $K =: K_{i+1}$ with $K_{i+1} \supset K^*_{l_{i+1}}$. This shows $\bigcup_{i \in \mathbb{N}} K_i = A(x_0)$.

For examples of the stepwise exhaustion, cf. Section 6.3 and Figure 5.4 as well as [23].

6

Application of the Method: Examples

In this chapter we summarize our method and apply it to several examples. We present three different construction methods for a Lyapunov function, which are illustrated by examples. For the data of the figures, cf. Appendix B.2.

1. Section 6.1: Combination of a local Lyapunov function $\mathfrak{d}$ or $\mathfrak{v}$ and a non-local approximation of T or V, cf. Sections 4.2.1 or 4.2.2.
2. Section 6.2: Approximation of V via W using the Taylor polynomial, cf. Section 4.2.3.
3. Section 6.3: Stepwise exhaustion using mixed approximation, cf. Section 5.2.

We conclude this chapter in Section 6.4 discussing several computational and conceptual aspects of the method.

For the grid we use in general the hexagonal grid in $\mathbb{R}^2$ and its generalizations in higher dimensions. Since we can use any grid, we are free to add points to the hexagonal grid where the orbital derivative has positive sign. Note that we know the values $Q'(x)$ where Q is the approximated function T or V for all points x.

The reason for the hexagonal grid is the following: on the one hand we are interested in a dense grid, in order to obtain a small fill distance h and thus, by the error estimate, a small error. On the other hand, the closer two points of the grid are, the worse is the condition number of the interpolation matrix A: if two points are equal, then the matrix is singular. Hence, we are seeking for the optimal grid with respect to these two opposite conditions: it turns out that this is the hexagonal grid, cf. [39]. The dilemma of these two contrary goals is also called the uncertainty relation, cf. [53] or [40]. The grid points of the hexagonal grid and its generalization in $\mathbb{R}^n$ are given by

$$\left\{\alpha \sum_{k=1}^{n} i_k w_k \mid i_k \in \mathbb{Z}\right\},$$

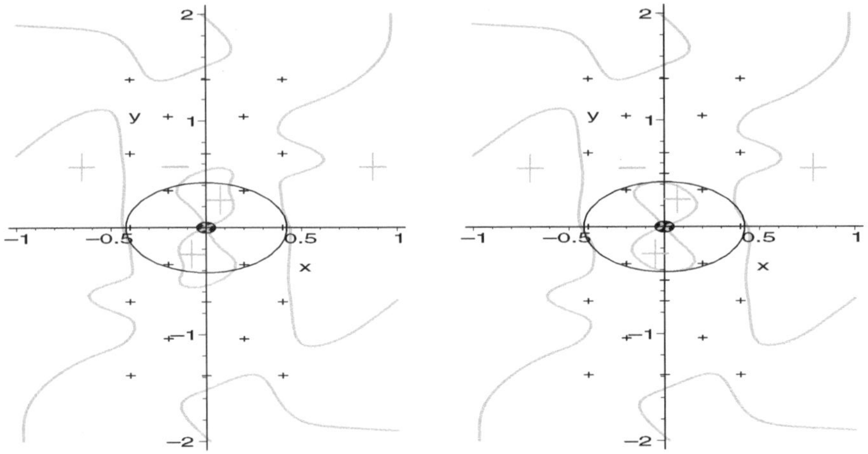

Fig. 6.1. Left: the grid (black +) and the sign of the approximation $v'(x, y)$ (grey). Obviously, there are points outside the local Lyapunov basin (thin black) such that $v'(x, y) > 0$. Right: we use a denser grid with two more points on the y-axis (black +). Now the sign of $v'(x, y)$ (grey) is only positive inside the local Lyapunov basin (thin black). We approximate the function V with $V'(x, y) = -(x^2 + y^2)$ of example (2.11).

$$\begin{aligned} \text{where } w_1 &= (2e_1, 0, 0, \ldots, 0) \\ w_2 &= (e_1, 3e_2, 0, \ldots, 0) \\ w_3 &= (e_1, e_2, 4e_3, 0, \ldots, 0) \\ &\vdots \\ w_n &= (e_1, \ldots, e_{n-1}, (n+1)e_n) \\ \text{with } e_k &= \sqrt{\frac{1}{2k(k+1)}}. \end{aligned}$$

The quantity $\alpha \in \mathbb{R}^+$ is proportional to the fill distance of the grid. For example, for dimension $n = 2$ we have the two vectors $w_1 = (1, 0)$ and $w_2 = \left(\frac{1}{2}, \frac{\sqrt{3}}{2}\right)$. For dimension $n = 3$ we have the three vectors $w_1 = (1, 0, 0)$, $w_2 = \left(\frac{1}{2}, \frac{\sqrt{3}}{2}, 0\right)$ and $w_3 = \left(\frac{1}{2}, \frac{1}{2\sqrt{3}}, \sqrt{\frac{2}{3}}\right)$.

We illustrate the use of a hexagonal grid and how to add points in a suitable way by example (2.11): In Figure 6.1, left, a hexagonal grid was used; however, there are points x with $v'(x) > 0$ outside the local Lyapunov basin. Thus, we add two points in these regions to the grid and with the denser grid we obtain an approximation v in Figure 6.1, right, for which the points with $v'(x) > 0$ lie inside the local Lyapunov basin. We have shown, using error estimates, that this is the expected behavior provided that the grid is dense enough.

6.1 Combination of a Local and Non-Local Lyapunov Function

6.1.1 Description

Consider $\dot{x} = f(x)$ with $f \in C^\sigma(\mathbb{R}^n, \mathbb{R}^n)$ and an exponentially asymptotically stable equilibrium x_0.

1. Local part
   - Calculate the local Lyapunov function $\mathfrak{q} = \mathfrak{d}$ or $\mathfrak{q} = \mathfrak{v}$.
   - Determine the set $\{x \in \mathbb{R}^n \mid \mathfrak{q}'(x) < 0\}$.
   - Find $r > 0$ such that $\tilde{K} := \tilde{K}_r^{\mathfrak{q}}(x_0) = \{x \in \mathbb{R}^n \mid \mathfrak{q}(x) \leq r^2\} \subset \{x \in \mathbb{R}^n \mid \mathfrak{q}'(x) < 0\} \cup \{x_0\}$ (local Lyapunov basin).
2. Preparation
   - Choose the radial basis function $\psi(r) = \psi_{l,k}(cr)$ with suitable $c > 0$ and $k \in \mathbb{N}$ such that $\mathbb{N} \ni \sigma \geq \sigma^* := \frac{n+1}{2} + k$, let $l = \lfloor \frac{n}{2} \rfloor + k + 1$.
   - Choose a grid X_N including no equilibrium.
3. Non-local part
   - Calculate the approximant q of $Q = T$ or $Q = V$ by solving $A\beta = \alpha$, for A cf. Proposition 3.5, $\alpha_j = -\bar{c}$ or $\alpha_j = -p(x_j)$.
   - Determine the set $\{x \in \mathbb{R}^n \mid q'(x) < 0\}$.
   - Find $R \in \mathbb{R}$ such that $K = \{x \in B \mid q(x) \leq R\} \subset \{x \in \mathbb{R}^n \mid q'(x) < 0\} \cup \overset{\circ}{\tilde{K}}$, where B is an open neighborhood of K.

Then $K \subset A(x_0)$ by Theorem 2.26, where $E = \overset{\circ}{\tilde{K}}$.

6.1.2 Examples

Example 6.1 (Speed-control) *As an example let us consider the system*

$$\begin{cases} \dot{x} = y \\ \dot{y} = -K_d\, y - x - gx^2 \left(\frac{y}{K_d} + x + 1\right) \end{cases} \tag{6.1}$$

with $K_d = 1$ and $g = 6$. This is a speed-control problem discussed in [16], [41] and [14]; this method has been applied to this example in [22]. The system (6.1) has the two asymptotically stable equilibria $x_0 = (0,0)$ and $(-0.7887, 0)$, and the saddle $(-0.2113, 0)$. The system fails to reach the demanded speed which corresponds to the equilibrium $(0,0)$ for some input since the basin of attraction of $x_0 = (0,0)$ is not the whole phase space, e.g. the unstable equilibrium $(-0.2113, 0)$ does not belong to the basin of attraction.

The eigenvalues of the matrix $Df(0,0) = \begin{pmatrix} 0 & 1 \\ -1 & -1 \end{pmatrix}$ are $\lambda_{1,2} = -\frac{1}{2} \pm \frac{\sqrt{3}}{2} i$. The eigenvectors are $v_{1,2} = \begin{pmatrix} 1 \\ \lambda_{1,2} \end{pmatrix}$. Following Lemmas 2.27 and 2.28 we

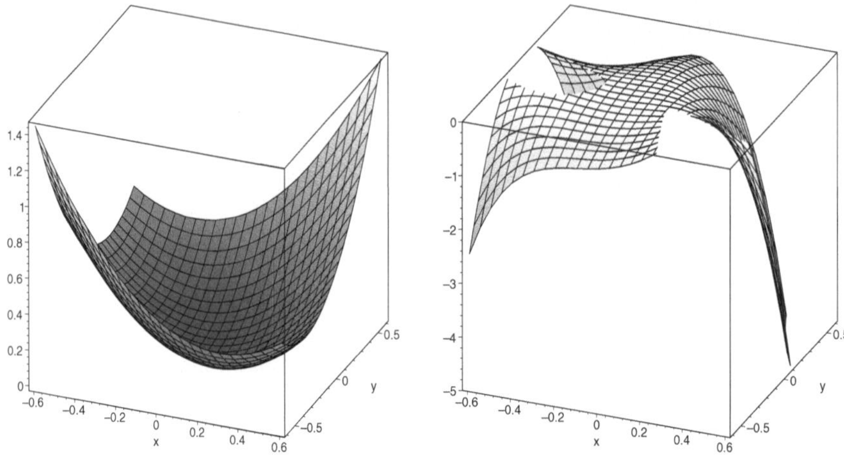

Fig. 6.2. Left: the function $\mathfrak{d}(x, y)$, which is a quadratic form. Right: the negative values of $\mathfrak{d}'(x, y)$ for (6.1).

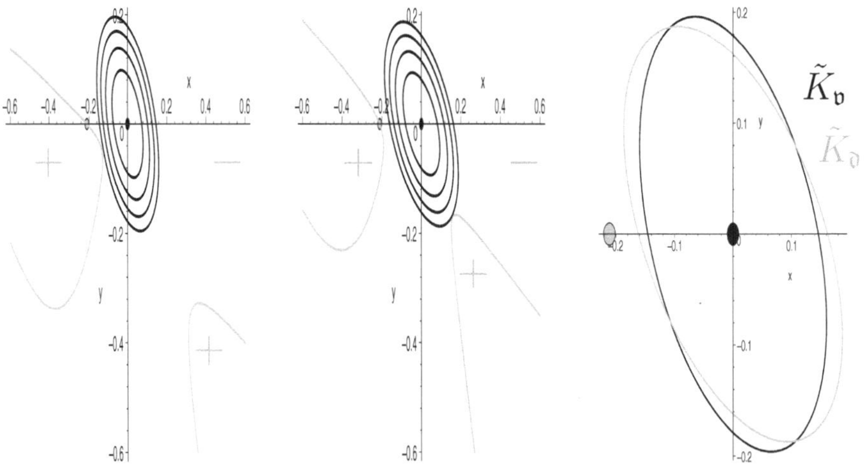

Fig. 6.3. Left: the sign of $\mathfrak{v}'(x, y)$ (grey) and several level sets of $\mathfrak{v}$ (black). Middle: the sign of $\mathfrak{d}'(x, y)$ (grey) and several level sets of $\mathfrak{d}$ (black). Right: comparison of the local Lyapunov basins $\tilde{K}_{\mathfrak{v}}$ (black) and $\tilde{K}_{\mathfrak{d}}$ (grey) for (6.1).

define the matrix $S = \begin{pmatrix} 1 & 0 \\ -\frac{1}{2} & \frac{\sqrt{3}}{2} \end{pmatrix}^{-1}$ *and* $\mathfrak{d}(x) = \|Sx\|^2$. *The function* $\mathfrak{v}$ *is obtained by solving the matrix equation* $Df(0,0)^T B + BDf(0,0) = -I$, *cf. Remark 2.34. With the solution* $B = \begin{pmatrix} \frac{3}{2} & \frac{1}{2} \\ \frac{1}{2} & 1 \end{pmatrix}$ *we set* $\mathfrak{v}(x) = x^T Bx$.

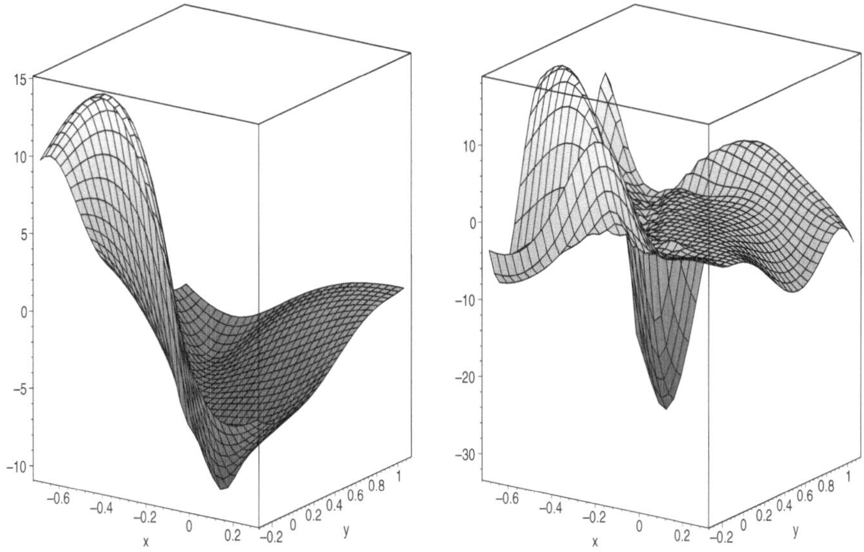

Fig. 6.4. Left: the function $t(x, y)$, approximating T with $T'(x, y) = -1$. Right: $t'(x, y)$ for (6.1).

Figure 6.2 shows the quadratic form $\mathfrak{d}(x)$ *and the negative values of its orbital derivative* $\mathfrak{d}'(x)$. *In 1. Local part we determine the level set* $\mathfrak{d}'(x) = 0$ *and, since* $\mathfrak{d}'$ *is a continuous function, we can determine the sign of* $\mathfrak{d}'$. *In Figure 6.3, middle, level sets* $\partial \tilde{K}_r^{\mathfrak{d}}(x_0)$ *for different* $r > 0$ *are plotted – since* $\mathfrak{d}$ *is a quadratic form these level sets are ellipses. One of the largest ellipses inside* $\{x \in \mathbb{R}^2 \mid \mathfrak{d}'(x) < 0\} \cup \{x_0\}$ *is denoted by* $\tilde{K}_{\mathfrak{d}}$ *and is a Lyapunov basin. The same is done for the quadratic form* $\mathfrak{v}$ *in Figure 6.3, left, and the Lyapunov basins* $\tilde{K}_{\mathfrak{d}}$ *and* $\tilde{K}_{\mathfrak{v}}$ *are compared in Figure 6.3, right.*

In 2. Preparation we choose the Wendland function with $k = 1$, $l = 3$, $c = \frac{5}{6}$ *and a hexagonal grid with* $\alpha = 0.05$ *and* $N = 223$ *points; for the data of all figures, cf. Appendix B.2. In 3. Non-local part we approximate the global Lyapunov function* T *with* $T'(x) = -1$, *i.e.* $\bar{c} = 1$, *by* t. *Figure 6.4 shows the approximating function* $t(x)$ *and its orbital derivative* $t'(x)$. *Note that* $t'(x) \approx -1$ *in the part of* $\mathbb{R}^2$ *where the grid points have been placed. The level set* $t'(x) = 0$ *is calculated and, since* t' *is a continuous function, it divides the regions with positive and negative sign of* t', *cf. Figure 6.5, right. There is a small region near* x_0 *where* $t'(x) > 0$, *this region, however, is a subset of the local Lyapunov basin* $\tilde{K}_{\mathfrak{v}}$. *We fix a value* $R \in \mathbb{R}$ *such that all points of* $K = \{x \in B \mid t(x) \leq R\}$ *either satisfy* $t'(x) < 0$ *or* $x \in \overset{\circ}{\tilde{K}}_{\mathfrak{v}}$. *By Theorem 2.26* K *is a subset of* $A(x_0)$. *Note that* $\tilde{K}_{\mathfrak{v}} \not\subset K$, *but this is not necessary for the application of Theorem 2.26.*

In order to combine the local and non-local function to a Lyapunov function using the Extension Theorem 4.8, it is necessary that $\tilde{K}_{\mathfrak{v}} \subset K$ *holds, which*

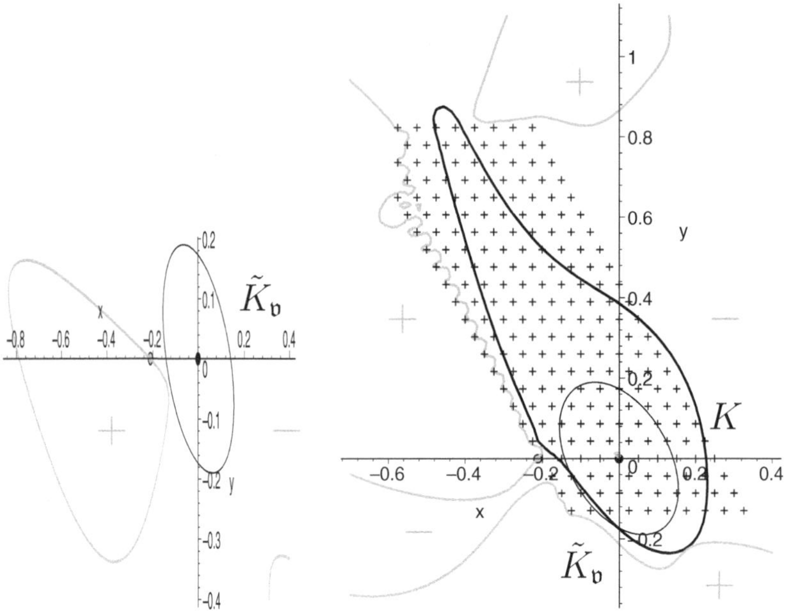

Fig. 6.5. Left: the sign of $\mathfrak{v}'(x, y)$ (grey) and the local Lyapunov basin $\tilde{K}_{\mathfrak{v}}$, the boundary of which is a level set of $\mathfrak{v}$ (thin black). Right: the sign of $t'(x, y)$ (grey), the grid points (black +), the set K, the boundary of which is a level set of t (black), and the local Lyapunov basin $\tilde{K}_{\mathfrak{v}}$ (thin black) for (6.1). K is a subset of the basin of attraction of the origin.

is not the case in this example. However, one could choose a smaller sublevel set $\tilde{K}_{\mathfrak{v}} = \tilde{K}_{\tilde{r}}^{\mathfrak{v}}(x_0)$ *with* $\tilde{r} < r$ *such that* $\tilde{K}_{\mathfrak{v}} \subset K$ *and* $t'(x) < 0$ *holds for all* $x \in K \setminus \tilde{K}_{\mathfrak{v}}$. *Then we can also proceed as in Section 4.2.2.*

Example 6.2 *As a second example we consider*

$$\begin{cases} \dot{x} = -x + x^3 \\ \dot{y} = -\frac{1}{2}y + x^2. \end{cases} \tag{6.2}$$

This example was presented at the NOLCOS 2004 IFAC meeting, cf. [22]. The equilibria of (6.2) are $(0, 0)$ *(asymptotically stable) and the two saddles* $(\pm 1, 2)$. *For this system we can determine the basin of attraction of* $(0, 0)$ *directly:* $A(0, 0) = \{(x, y) \in \mathbb{R}^2 \mid -1 < x < 1\}$. *Thus, we can compare our calculations with the exact basin of attraction.*

For 1. Local part we use the local Lyapunov function $\mathfrak{v}$, *where* $\mathfrak{v}(x, y) = (x, y)B\begin{pmatrix} x \\ y \end{pmatrix}$ *with* $B = \begin{pmatrix} \frac{1}{2} & 0 \\ 0 & \frac{1}{4} \end{pmatrix}$.

In 2. Preparation, we choose $k = 1$, $l = 3$ *and* $c = \frac{1}{2}$. *Our hexagonal grid has* $N = 122$ *points,* $\alpha = 0.3$, *and we approximate* V_1 *where* $V_1'(x) = -\|x\|^2$.

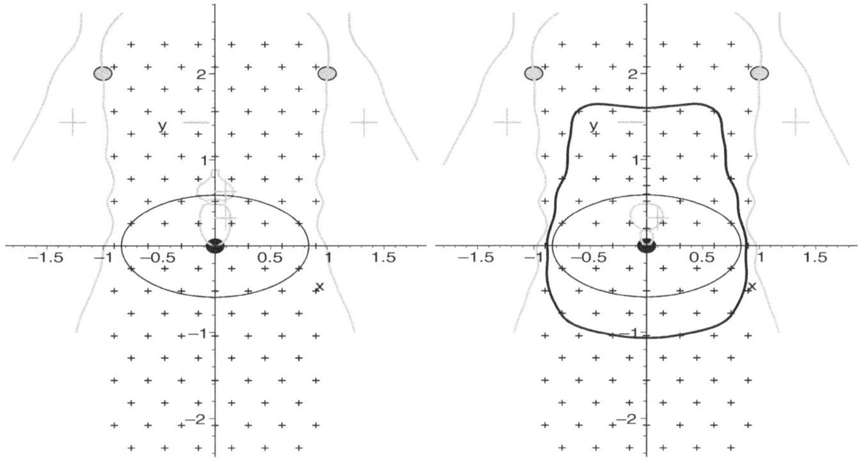

Fig. 6.6. Left: the local Lyapunov basin $\tilde{K}_{\mathfrak{v}}$ (thin black), the boundary of which is a level set of $\mathfrak{v}$, and the sign of v_1' (grey), where v_1 is an approximation of V_1 with $V_1'(x,y) = -(x^2+y^2)$ using the grid points (black +); there is a region outside $\tilde{K}_{\mathfrak{v}}$ where $v_1'(x,y) > 0$. Right: here we used the same grid plus four additional grid points on the y-axis (black +). Now the sign of $v_1'(x,y)$ (grey) is negative outside $\tilde{K}_{\mathfrak{v}}$ (thin black). There is a set K (black), the boundary of which is a level set of v_1, which is a Lyapunov basin for (6.2).

In Figure 6.6, left, there are points with $v_1'(x) > 0$ which do not lie inside the local Lyapunov basin $\tilde{K}_{\mathfrak{v}}$. Thus, we cannot apply Theorem 2.24 nor Theorem 2.26.

Hence, we use a denser grid. It turns out that adding four additional points on the y-axis suffices. Now we have $N = 126$ grid points, and the conditions are satisfied. In Figure 6.6, right, we can proceed with 3. Non-local part. The level set $v_1'(x) = 0$ is calculated. There is a small region near x_0 where $v_1'(x) > 0$; this region, however, is now a subset of the local Lyapunov basin $\tilde{K}_{\mathfrak{v}}$. We fix a value $R \in \mathbb{R}$ such that all points of $K = \{x \in B \mid v_1(x) \leq R\}$ either satisfy $v_1'(x) < 0$ or $x \in \mathring{\tilde{K}}_{\mathfrak{v}}$. By Theorem 2.26, K is a subset of $A(x_0)$.

Moreover, we compare the approximations v_1, v_2 and t of the functions

- $V_1'(x) = -\|x\|^2$
- $V_2'(x) = -\|f(x)\|^2$
- $T'(x) = -1$

in Figure 6.7. For all three approximants there are small regions near x_0 where $v_1'(x), v_2'(x), t'(x) > 0$, however, they lie inside $\tilde{K}_{\mathfrak{v}}$. In the last two cases this is already obtained for the grid with $N = 122$ points.

In all three examples of (6.2), cf. Figure 6.7, the local Lyapunov basin is a subset of the calculated Lyapunov basin, i.e. $\tilde{K}_{\mathfrak{v}} \subset K$. Thus, we can apply the results of Section 4.2.2 using the extension of the approximated functions

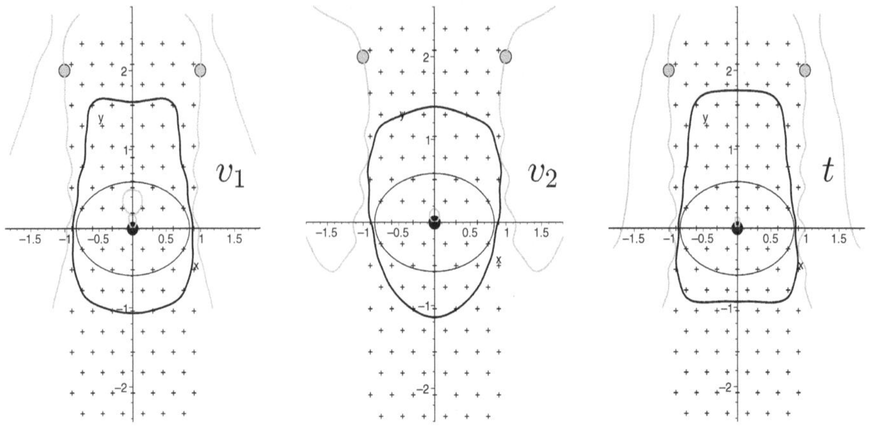

Fig. 6.7. All three figures show the grid (black +), the local Lyapunov basin $\tilde{K}_{\mathfrak{v}}$ (thin black), the set $q'(x, y) = 0$ (grey) and a Lyapunov basin, which is bounded by a level set of q (black) for (6.2). Left: $q = v_1$, where v_1 is the approximation of V_1 with $V_1'(x, y) = -(x^2 + y^2)$; here we used a grid with $N = 126$ points, cf. Figure 6.6, right. Middle: $q = v_2$, where v_2 is the approximation of V_2 with $V_2'(x, y) = -\|f(x, y)\|^2$; here we used a grid with $N = 122$ points. Right: $q = t$, where t is the approximation of T with $T'(x, y) = -1$; here we used a grid with $N = 122$ points.

v_1, v_2, t, respectively, and use Theorem 2.24. For (6.1), Figure 6.5, however, the local Lyapunov basin is not a subset of the calculated one, and thus we have to use Theorem 2.26.

6.2 Approximation via Taylor Polynomial

6.2.1 Description

Consider $\dot{x} = f(x)$ with $f \in C^\sigma(\mathbb{R}^n, \mathbb{R}^n)$ and an exponentially asymptotically stable equilibrium x_0. Fix the parameter of the Wendland function $k \in \mathbb{N}$ and let $\sigma \geq P \geq 2 + \sigma^*$, where $\sigma^* := \frac{n+1}{2} + k$.

1. Calculation of $\mathfrak{n}$
   - Calculate the Taylor polynomial $\mathfrak{h}$ of V with $V'(x) = -\|x - x_0\|^2$ of order P solving $\langle \nabla \mathfrak{h}(x), f(x) \rangle = -\|x - x_0\|^2 + o(\|x - x_0\|^P)$, cf. Definition 2.52 and Remark 2.54.
   - Find a constant $M \geq 0$ such that $\mathfrak{n}(x) = \mathfrak{h}(x) + M\|x - x_0\|^{2H} > 0$ holds for all $x \neq x_0$, cf. Definition 2.56.
2. Preparation
   - Choose the radial basis function $\psi(r) = \psi_{l,k}(cr)$ with suitable $c > 0$, let $l = \left\lfloor \frac{n}{2} \right\rfloor + k + 1$, where k was fixed above.
   - Choose a grid X_N including no equilibrium.

3. Approximation
   - Calculate the approximant $w(x)$ of $W(x) = \frac{V(x)}{\mathfrak{n}(x)}$ by solving $A\beta = \alpha$, where $m(x) = \frac{\mathfrak{n}'(x)}{\mathfrak{n}(x)}$ cf. Proposition 3.6 and $\alpha_j = -\frac{\|x_j - x_0\|^2}{\mathfrak{n}(x_j)}$. Set $v_W(x) = w(x)\,\mathfrak{n}(x)$.
   - Determine the set $\{x \in \mathbb{R}^n \mid v_W'(x) < 0\}$.
   - Find $R \in \mathbb{R}$ such that $K = \{x \in B \mid v_W(x) \leq R\} \subset \{x \in \mathbb{R}^n \mid v_W'(x) < 0\} \cup \{x_0\}$, where B is an open neighborhood of K.

Then $K \subset A(x_0)$ by Theorem 2.24.

Note that we can use Theorem 2.24 also for negative sublevel sets $K = \{x \in B \mid v_W(x) \leq R\}$ with $R \in \mathbb{R}_0^-$. Indeed, apply Theorem 2.24 to the function $\tilde{v}_W(x) := v_W(x) + C$ where $C > -\min_{y \in K} v_W(y)$. Then $K = \{x \in B \mid \tilde{v}_W(x) \leq R + C\}$ and $R + C \in \mathbb{R}^+$.

6.2.2 Examples

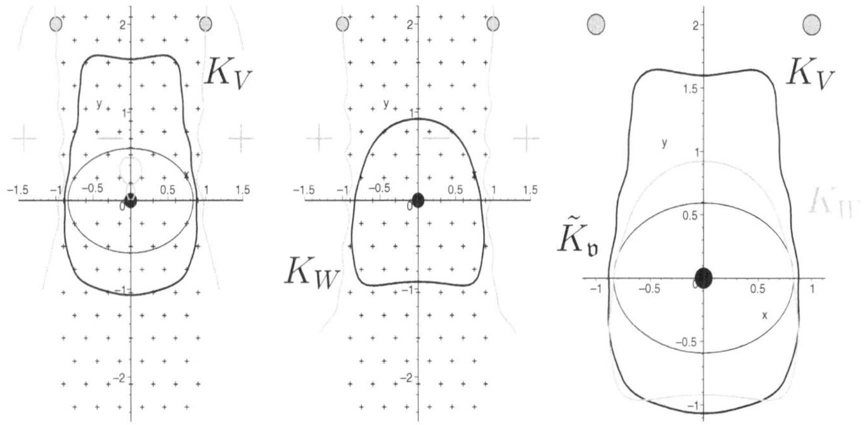

Fig. 6.8. All three figures show Lyapunov basins for (6.2). Left: the grid with $N = 126$ points (black +), the local Lyapunov basin $\tilde{K}_{\mathfrak{v}}$ (thin black), the sign of v_1' (grey) and a Lyapunov basin K_V (black) for v_1, which is bounded by a level set of v_1, where v_1 is the approximation of V_1 with $V_1'(x,y) = -(x^2 + y^2)$, cf. Example 6.2 and Figure 6.7, left. Middle: a grid with $N = 122$ points (black +), the sign of v_W' (grey) and a Lyapunov basin K_W (black) for v_W, which is bounded by a level set of v_W, where $v_W(x,y) = w(x,y)\,\mathfrak{n}(x,y)$ and w is the approximation of W for (6.2); this is the approximation of V using the Taylor polynomial. Right: a comparison of the three Lyapunov basins $\tilde{K}_{\mathfrak{v}}$ (local, thin black), K_V (direct approximation of V, black) and K_W (approximation of V via W using the Taylor polynomial, grey).

Example 6.3 *As an example consider the system (6.2) again, cf. Example 6.2. We have $n = 2$ and fix $k = 1$, thus $\sigma^* = \frac{5}{2}$. For 1. Calculation of $\mathfrak{n}$ we* *first calculate the Taylor polynomial* $\mathfrak{h}$ *of* V *for* $P = 5$, *cf. (2.31) of Example 2.55:*

$$\mathfrak{h}(x, y) = \underbrace{\frac{1}{2}x^2 + y^2}_{=\mathfrak{v}(x,y)} + \frac{4}{5}x^2y + \frac{9}{20}x^4 + \frac{16}{45}x^4y.$$

For the function $\mathfrak{n}$ *we have, cf. Example 2.57:*

$$\begin{aligned}\mathfrak{n}(x, y) &= \mathfrak{h}(x) + (x^2 + y^2)^3 \\ &= \frac{1}{2}x^2 + y^2 + \frac{4}{5}x^2y + \frac{9}{20}x^4 + \frac{16}{45}x^4y + (x^2 + y^2)^3.\end{aligned}$$

Now we approximate $W(x) = \frac{V(x)}{\mathfrak{n}(x)}$, *where* $V'(x) = -\|x\|^2$, *by* w. *For the function* $v_W(x) = w(x)\,\mathfrak{n}(x)$ *we determine the set* $v_W'(x) = 0$, *which divides the region of positive and negative sign of the continuous function* v_W'. *As shown in Figure 6.8, middle,* $v_W'(x) < 0$ *holds for all* x *near* x_0. *We find a Lyapunov basin* K_W, *cf. Figure 6.8, middle.*

We compare this Lyapunov basin K_W *with the Lyapunov basin* K_V *for the approximation* v_1 *of* $V_1 = V$ *from Example 6.2 in Figure 6.8, left and right.*

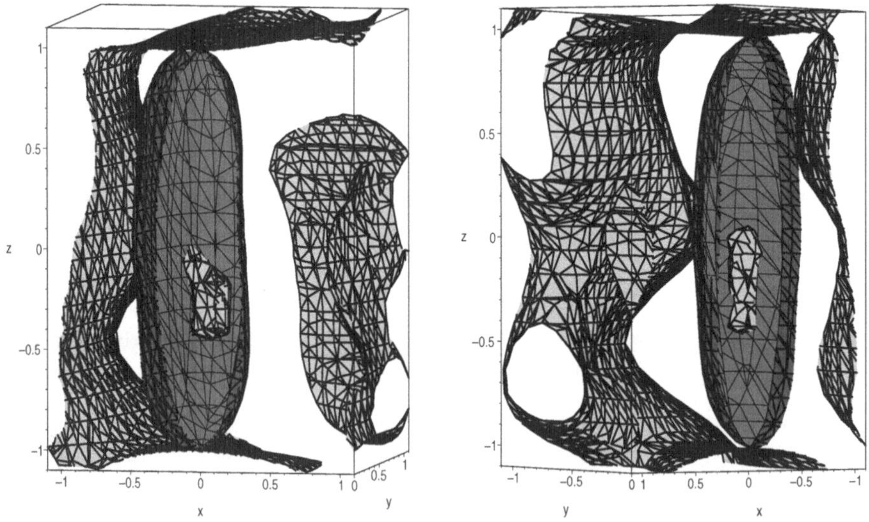

Fig. 6.9. Direct approximation of V. The figures show the local Lyapunov basin $\tilde{K}_{\mathfrak{v}}$ (dark grey), the set $v'(x, y, z) = 0$ (grey) and a Lyapunov basin K_V (black) for v, which is bounded by a level set of v for (6.3). Near the equilibrium $(0, 0, 0)$ there is a set where $v'(x, y, z)$ is positive. Left: the points $(x, y, z) \in \mathbb{R}^3$ with $y > 0$. Right: the points $(x, y, z) \in \mathbb{R}^3$ with $y < 0$.

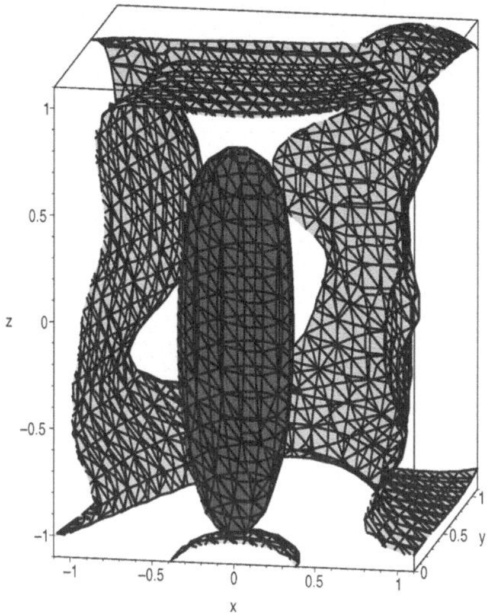

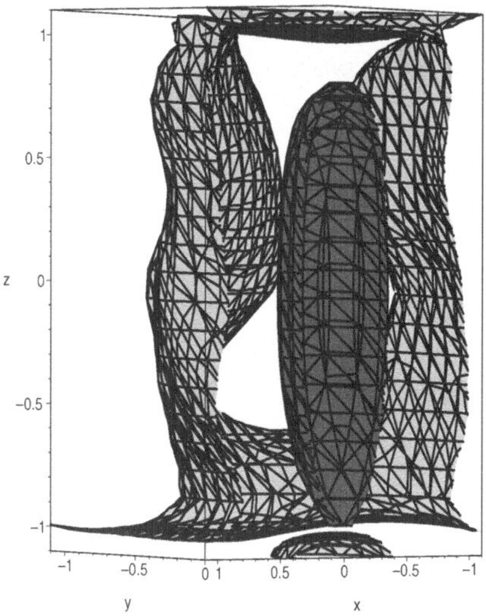

Fig. 6.10. Approximation of V via W using the Taylor polynomial. The figures show the set $v'_W(x,y,z) = 0$ (grey) and a Lyapunov basin K_W (black) for v_W, which is bounded by a level set of v_W for (6.3). Near the equilibrium $(0,0,0)$ the function $v'(x,y,z)$ is negative. Left: the points $(x,y,z) \in \mathbb{R}^3$ with $y > 0$. Right: the points $(x,y,z) \in \mathbb{R}^3$ with $y < 0$.

Example 6.4 (A three-dimensional example) *We consider the system*

$$\begin{cases} \dot{x} = x(x^2 + y^2 - 1) - y(z^2 + 1) \\ \dot{y} = y(x^2 + y^2 - 1) + x(z^2 + 1) \\ \dot{z} = 10z(z^2 - 1). \end{cases} \tag{6.3}$$

The basin of attraction of the asymptotically stable equilibrium $(0,0,0)$ *is given by* $A(0,0,0) = \{(x,y,z) \in \mathbb{R}^3 \mid x^2 + y^2 < 1, |z| < 1\}$.

We use $k = 1$, $l = 3$ *and* $c = 0.45$. *We approximate* V *with* $V'(x) = -\|x\|^2$ *directly (cf. Section 6.1) and via* W *using the Taylor polynomial (this section). We use the same hexagonal grid with* $N = 137$ *points and* $\alpha = 0.35$ *for both approximations.*

With $P = 5$ *we calculate the function* $\mathfrak{h}(x,y,z) = \frac{1}{2}x^2 + \frac{1}{2}y^2 + \frac{1}{20}z^2 + \frac{1}{4}x^4 + \frac{1}{4}y^4 + \frac{1}{40}z^4 + \frac{1}{2}x^2y^2$. *One immediately sees that* $\mathfrak{h}(x,y,z) > 0$ *holds for all* $(x,y,z) \neq (0,0,0)$. *Thus we set* $\mathfrak{n}(x,y,z) = \mathfrak{h}(x,y,z)$.

For the figures note that we split the three-dimensional figures in the positive and negative y-axis. We denote by K_V *the Lyapunov basin obtained by the direct approximation of* V, *cf. Figure 6.9 – here again we find a set with* $v'(x,y,z) > 0$ *near* $(0,0,0)$ *which lies inside the local Lyapunov basin* $\tilde{K}_{\mathfrak{v}}$.

We denote by K_W the Lyapunov basin obtained by the approximation of V via W using the Taylor polynomial, cf. Figure 6.10. In this case all the points (x, y, z) near $(0, 0, 0)$ satisfy $v'_W(x, y, z) < 0$.

6.3 Stepwise Exhaustion Using Mixed Approximation

6.3.1 Description

Consider $\dot{x} = f(x)$ with $f \in C^\sigma(\mathbb{R}^n, \mathbb{R}^n)$ and an exponentially asymptotically stable equilibrium x_0. Let $k \in \mathbb{N}$ and $\sigma \geq \sigma^*$, where $\sigma^* := \frac{n+1}{2} + k$, and let $l = \left\lfloor \frac{n}{2} \right\rfloor + k + 1$.

1. Step 0: local part
   - Calculate the local Lyapunov function $q_0 = \mathfrak{d}$ or $q_0 = \mathfrak{v}$.
   - Determine the set $\{x \in \mathbb{R}^n \mid q'_Q(x) < 0\}$.
   - Find $r_0 > 0$ such that $K_0 := \tilde{K}^{q_0}_{r_0}(x_0) = \{x \in \mathbb{R}^n \mid q_0(x) \leq r_0^2\} \subset \{x \in \mathbb{R}^n \mid q'_0(x) < 0\} \cup \{x_0\}$.
2. Step $i + 1 \geq 1$
   - Choose the radial basis function $\psi(r) = \psi_{l,k}(cr)$ with $c > 0$.
   - Choose a grid $X^0_M \subset \partial K_i$.
   - Choose a grid $X_N \subset \mathbb{R}^n \setminus \tilde{K}^{q_i}_{r_i - \epsilon}(x_0)$ with some $\epsilon > 0$.
   - Calculate the approximant $q_{i+1}(x)$ of $Q = T$ or $Q = V^*$ by solving $\begin{pmatrix} A & C \\ C^T & A^0 \end{pmatrix} \begin{pmatrix} \beta \\ \gamma \end{pmatrix} = \begin{pmatrix} \alpha \\ \alpha^0 \end{pmatrix}$, cf. Proposition 3.8, $\alpha_j = -\bar{c}$ or $\alpha_j = -p(x_j)$ and $\alpha^0_j = 1$.
   - Determine the set $\{x \in \mathbb{R}^n \mid q'_{i+1}(x) < 0\}$.
   - Find $r_{i+1} > 1$ such that $K_{i+1} = \{x \in B \mid q_{i+1}(x) \leq r_{i+1}^2\} \subset \{x \in \mathbb{R}^n \mid q'_{i+1}(x) < 0\} \cup \overset{\circ}{K}_i$, where B is an open neighborhood of K_{i+1}.

Then $K_{i+1} \subset A(x_0)$ by Theorem 2.26, where $E = \overset{\circ}{K}_i$.

6.3.2 Example

Example 6.5 (van-der-Pol oscillator) *As an example consider the system*

$$\begin{cases} \dot{x} = -y \\ \dot{y} = x - 3(1 - x^2)y. \end{cases} \tag{6.4}$$

This is the van-der-Pol system with inverse time. Hence, $(0, 0)$ is an asymptotically stable equilibrium and the boundary of its basin of attraction is a periodic orbit.

We calculate the function $\mathfrak{v}(x) = x^T B x$ by solving the matrix equation $Df(0, 0)^T B + BDf(0, 0) = -I$, the solution is $B = \begin{pmatrix} \frac{11}{6} & -\frac{1}{2} \\ -\frac{1}{2} & \frac{1}{3} \end{pmatrix}$. We obtain a local Lyapunov basin $\tilde{K} =: K_0$.

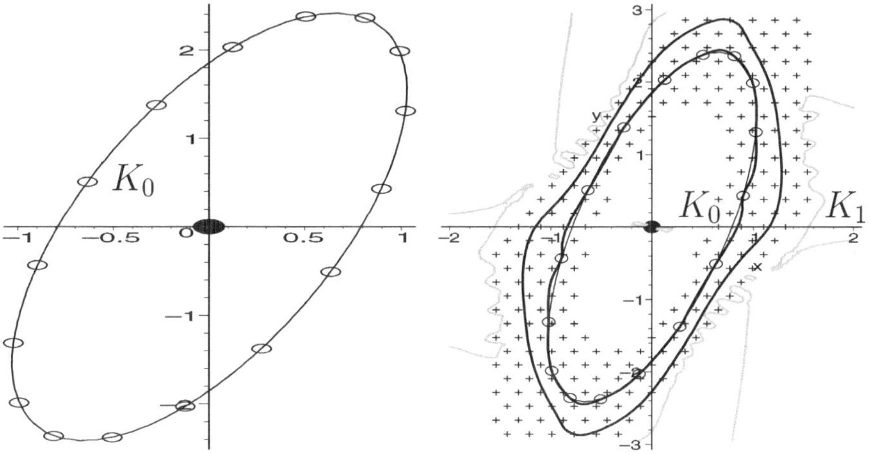

Fig. 6.11. Left: a local Lyapunov basin $\tilde{K} = K_0$ (thin black) and the grid points of X_M^0 (black circles). Right: the grid X_N (black +), the grid X_M^0 (black circles), the local Lyapunov basin K_0 (thin black), the set $v'(x, y) = 0$ (grey), where v is the mixed approximation of V^* with $(V^*)'(x, y) = -(x^2 + y^2)$, and the level sets $v(x, y) = 1$ and $v(x, y) = 1.6$ (black) for (6.4). The set K_1 (black) is the sublevel set $v(x, y) \leq 1.6$.

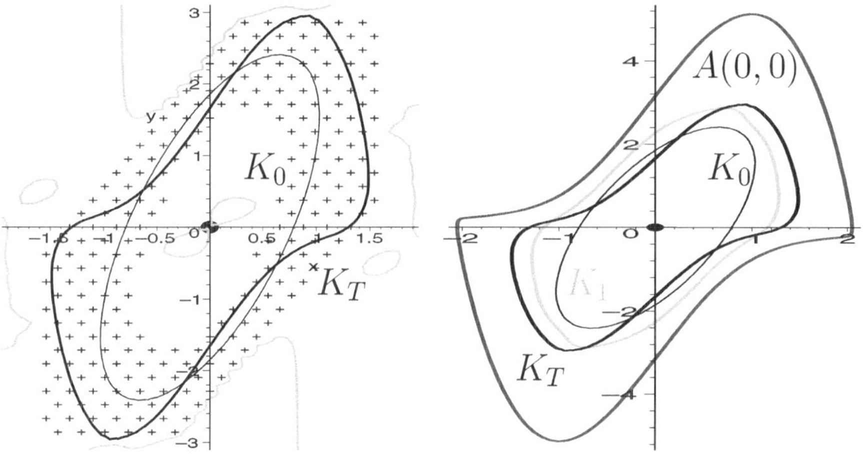

Fig. 6.12. Left: a local Lyapunov basin $\tilde{K} = K_0$ (thin black), the grid X_N (black +), the set $t'(x, y) = 0$ (grey), where t is the approximation of T with $T'(x, y) = -2.5$ only via the orbital derivative, a level set K_T of t (black) for (6.4). Here we can apply Theorem 2.26 but we cannot enlarge the Lyapunov basin. Right: comparison of the local Lyapunov basin K_0 (thin black), the Lyapunov basin obtained by mixed approximation K_1 (grey), cf. Figure 6.11, and by an approximation via the orbital derivative K_T (black) as well as the numerically calculated periodic orbit, i.e. the boundary of the basin of attraction $A(0, 0)$ (dark grey).

In the next step we use a hexagonal grid X_N of $N = 236$ points and $\alpha = 0.22$. The grid X_M^0 consists of $M = 16$ points on $\Omega = \partial\tilde{K}$, cf. Figure 6.11, left. We choose $c = \frac{10}{21}$ and approximate the function V^* with $(V^*)'(x) = -\|x\|^2$. The level set $v(x) = 1$ is approximately Ω. The set $K_1 = \{x \in B \mid v(x) \leq 1.6\}$ is a Lyapunov basin with $K_1 \supset K_0 = \tilde{K}$, cf. Figure 6.11, right.

If we do not use a mixed approximation but an approximation t via the orbital derivative as in Section 6.1, one cannot control the values of t. Figure 6.12 shows an approximation t of T with $T'(x) = -2.5$ with the same grid X_N only via the orbital derivative. The Lyapunov basin K_T is a subset of the basin of attraction with Theorem 2.26 and exceptional set $\overset{\circ}{\tilde{K}}$. However, since $K_T \not\supset \tilde{K}$, we cannot define a Lyapunov function by combining the local Lyapunov function with t. Thus, we cannot do the next step to enlarge the Lyapunov basin. Figure 6.12, right, compares the local Lyapunov basin K_0, the Lyapunov basin K_1 of the mixed approximation and the Lyapunov basin K_T of the approximation via the orbital derivative, respectively, with the numerically calculated periodic orbit, which is the boundary of the basin of attraction $A(0,0)$.

For an example of this method with more steps, cf. [23].

6.4 Conclusion

We use Lyapunov functions to determine the basin of attraction of an exponentially asymptotically stable equilibrium. Among all Lyapunov functions we have considered two classes: functions with a constant orbital derivative $T'(x) = -\bar{c}$ and functions, where the orbital derivative is a smooth function which tends to zero for $x \to x_0$; we denote the latter class of Lyapunov functions by V where $V'(x) = -p(x)$. The difference between these two classes is that (i) T is not defined in x_0, (ii) the existence proofs for T and V are different, (iii) the estimates for t' and v' are obtained differently and (iv) V is uniquely determined by the above property, whereas T is only unique if the values of $T(x)$ are fixed on a non-characteristic hypersurface.

We can estimate the approximation error for $Q = T$ and $Q = V$ by $|Q'(x) - q'(x)| \leq \iota$. Since $V'(x) = -p(x)$ which tends to 0 for $x \to x_0$, a small ι and thus a dense grid is needed near x_0. Far away from x_0, however, $p(x)$ is possibly very large and a small value of ι is not needed to ensure that $q'(x)$ is negative. Thus, for large basins of attraction, the approximation of T can be more appropriate, since here $t'(x) \leq T'(x) + \iota = -\bar{c} + \iota$ holds for all points x and hence a small ι is important for all x. In examples, however, the approximation of V and T often do not show very different results; a reason may be that the error estimate of the theorem is too restrictive and in applications the grid does not need to be as dense as required by the theorem.

When calculating a Lyapunov basin, the basin of attraction is not known beforehand. A natural problem is thus, in which part of $\mathbb{R}^n$ to choose the grid.

What happens if we choose grid points which do not belong to the basin of attraction? This case is not covered by the theorems, but in fact either q' is positive near these points, cf. Figure 6.5, or the level sets do not reach these points, cf. Figure 5.2. In both cases some points of the grid are not in the basin of attraction. However, the Lyapunov basin obtained is always a subset of the basin of attraction by Theorem 2.26, no matter where we have obtained the Lyapunov function from.

Thus, after choosing a grid we calculate a function q. If there are points x with $q'(x) > 0$ outside the local Lyapunov basin, then either these points do not belong to the basin of attraction or we have to add points to the grid, however, it is not possible to distinguish between both cases. The theorems only provide sufficient conditions for a set to belong to the basin of attraction, we cannot show that points do *not* belong to the basin of attraction.

Radial basis functions can be used for dynamical systems in more ways: Using radial basis functions to approximate Lyapunov functions can also be applied to discrete dynamical systems [25]. Radial basis functions can also be used for the determination of the basin of attraction of periodic orbits in continuous dynamical systems [26]. Moreover, the method of this book can be generalized to time-periodic differential equation, cf. [28].

In this book we provided a general method to construct a Lyapunov function for an exponentially asymptotically stable equilibrium of an autonomous ordinary differential equation. We considered Lyapunov functions satisfying a certain linear partial differential equation and proved their existence. We used radial basis functions to approximate solutions of these partial differential equations and thus to construct a Lyapunov function. We used the sublevel set of these approximative solutions to determine a subset of the basin of attraction of the equilibrium. We showed that we can cover each compact subset of the basin of attraction by a sublevel set of a constructed Lyapunov function. Hence, we provided a method to construct global Lyapunov functions.

A

Distributions and Fourier Transformation

In this appendix we give a summary of the facts of Distributions and Fourier transformation that we need in Chapter 3. We follow Hörmander [37], Chapter I., and Yosida [69], pp. 46–52. The proofs of the results of this section can be found in these books. In the appendix we consider complex-valued functions in general, i.e. $C^k(\mathbb{R}^n)$, $C^\infty(\mathbb{R}^n)$ denotes $C^k(\mathbb{R}^n, \mathbb{C})$, $C^\infty(\mathbb{R}^n, \mathbb{C})$.

A.1 Distributions

Definition A.1 (supp). *The support of* $\varphi \in C^k(\mathbb{R}^n)$*,* $k \geq 0$*, is defined as* $\text{supp}(\varphi) = \overline{\{x \in \mathbb{R}^n \mid \varphi(x) \neq 0\}}$.

Let $\Omega \subset \mathbb{R}^n$ *be a set.* $C_0^\infty(\Omega)$ *denotes the space of functions in* $C^\infty(\mathbb{R}^n)$ *with compact support which is contained in* Ω*.*

We define the space $\mathcal{D}'(\mathbb{R}^n)$ of distributions. The condition (A.1) is equivalent to the continuity of T with respect to a certain norm of $C_0^\infty(\mathbb{R}^n)$.

Definition A.2 (Distribution). *A linear operator* $T\colon C_0^\infty(\mathbb{R}^n) \to \mathbb{C}$ *is called a distribution, if for each compact set* $K \subset \mathbb{R}^n$ *there are constants* C *and* k*, such that*

$$|T(\varphi)| \leq C \sum_{|\alpha| \leq k} \sup_{x \in K} |D^\alpha \varphi(x)| \tag{A.1}$$

holds for all $\varphi \in C_0^\infty(K)$*.*

Here we define for the multiindex $\alpha = (\alpha_1, \alpha_2, \ldots, \alpha_n) \in \mathbb{N}_0^n$ *the following expressions:* $|\alpha| := \sum_{j=1}^n \alpha_j$ *and* $D^\alpha := \partial_{x_1}^{\alpha_1} \ldots \partial_{x_n}^{\alpha_n}$*. The space of all distributions is denoted by* $\mathcal{D}'(\mathbb{R}^n)$ *since it is the dual of the space* $\mathcal{D}(\mathbb{R}^n) := C_0^\infty(\mathbb{R}^n)$ *of test functions. We also write* $T(\varphi) = \langle T, \varphi \rangle$*.*

The last notation is derived from the L^2 scalar product. The next example shows that all locally integrable functions define a distribution.

Example A.3 *Every function $f \in L^1_{loc}(\mathbb{R}^n)$, i.e. f is locally integrable, defines a distribution T through*

$$T(\varphi) := \int_{\mathbb{R}^n} f(x)\varphi(x)\, dx = \langle f(x), \varphi(x)\rangle.$$

Indeed, we have $\left|\int_{\mathbb{R}^n} f(x)\varphi(x)\, dx\right| \leq \sup_{x\in K} |\varphi(x)| \int_K |f(x)|\, dx$ for a test function $\varphi \in C_0^\infty(K)$.

Thus, all L^1_{loc}-functions belong to $\mathcal{D}'(\mathbb{R}^n)$. But the space of distributions is larger than L^1_{loc}; e.g. Dirac's delta-distribution is not an L^1_{loc}-function.

Example A.4 *Dirac's δ-distribution δ is defined by $\langle \delta, \varphi\rangle := \varphi(0)$ for all $\varphi \in C_0^\infty(\mathbb{R}^n)$.*

δ is a distribution, but no L^1_{loc} function.

We show that δ is a distribution: For $0 \notin K$ this is clear. Now let $0 \in K$. For $x \in K$ and $\varphi \in C_0^\infty(K)$ we have $|\langle \delta, \varphi\rangle| = |\varphi(0)| \leq \sup_{x\in K} |\varphi(x)|$.

We show that δ is no L^1_{loc}-function. Let χ be a function with $\chi \in C_0^\infty(\mathbb{R}^n; [0,1])$, $\mathrm{supp}(\chi) \subset \overline{B_1(0)}$ and $\chi(x) = 1$ for $\|x\| \leq \frac{1}{2}$. Set $\varphi_\epsilon(x) := \chi\left(\frac{x}{\epsilon}\right)$. Hence, $\mathrm{supp}\,\varphi_\epsilon \subset \overline{B_\epsilon(0)}$. We assume that there is a function $f \in L^1_{loc}(\mathbb{R}^n)$ such that

$$1 = |\varphi_\epsilon(0)| = |\langle \delta, \varphi_\epsilon\rangle| = \left|\int_{\mathbb{R}^n} f(x)\varphi_\epsilon(x)\, dx\right| \leq \int_{\mathbb{R}^n} 1_{\overline{B_\epsilon(0)}}(x)|f(x)|\, dx.$$

By Lebesgue's Theorem this term tends to zero for $\epsilon \to 0$, since $1_{\overline{B_\epsilon(0)}}$ converges to zero almost everywhere and $\int_{\|x\|\leq 1} |f(x)|\, dx < \infty$ if $f \in L^1_{loc}(\mathbb{R}^n)$, contradiction. Thus, δ is no L^1_{loc}-function.

Definition A.5 (Support of a distribution). *Let $T \in \mathcal{D}'(\mathbb{R}^n)$, and let $\Omega \subset \mathbb{R}^n$ be an open set.*

1. *We say that $T = 0$ on Ω, if*

$$\langle T, \varphi\rangle = 0$$

   holds for all $\varphi \in C_0^\infty(\Omega)$.
2. *The* support of a distribution *T is the complement of the largest open set, where $T = 0$ holds.*
3. *The space of all distributions with compact support is denoted by $\mathcal{E}'(\mathbb{R}^n)$. It is the dual of the space $\mathcal{E}(\mathbb{R}^n) = C^\infty(\mathbb{R}^n)$.*

Example A.6 *The δ-distribution has the support $\{0\}$. Thus, it belongs to $\mathcal{E}'(\mathbb{R}^n)$ and one can apply it to functions of $C^\infty(\mathbb{R}^n)$.*

We now define operations on distributions. The idea is to apply the respective operation to the smooth test functions. We define multiplication by a smooth function, differentiation etc. Note, that for a function φ we define $\check{\varphi}(x) = \varphi(-x)$.

Definition A.7. *Let $T \in \mathcal{D}'(\mathbb{R}^n)$, $\varphi \in C_0^\infty(\mathbb{R}^n)$.*

1. Multiplication by a function $a \in C^\infty(\mathbb{R}^n)$*; we have* $aT \in \mathcal{D}'(\mathbb{R}^n)$.

$$\langle aT, \varphi \rangle := \langle T, a\varphi \rangle.$$

2. Conjugation

$$\langle \overline{T}, \varphi \rangle := \overline{\langle T, \overline{\varphi} \rangle}.$$

3. Check

$$\langle \check{T}, \varphi \rangle := \langle T, \check{\varphi} \rangle.$$

4. Differentiation*; we have* $\frac{\partial}{\partial x_j} T \in \mathcal{D}'(\mathbb{R}^n)$.

$$\left\langle \frac{\partial}{\partial x_j} T, \varphi \right\rangle := -\left\langle T, \frac{\partial}{\partial x_j} \varphi \right\rangle.$$

These formulas also hold if T is given by a smooth function f through $T(\varphi) = \int_{\mathbb{R}^n} f(x)\varphi(x)\,dx$. For 3. this is shown by partial integration; note, that the test functions have compact support.

Definition A.8 (Convolution). *For two continuous functions f, g, one of which has compact support, we define $(f * g)(x) := \int_{\mathbb{R}^n} f(y)g(x-y)\,dy$.*

Definition A.9 (Convolution for distributions). *We define the convolution of a distribution $T \in \mathcal{D}'(\mathbb{R}^n)$ with a function $\varphi \in C_0^\infty(\mathbb{R}^n)$ as follows:*

$$(T * \varphi)(x) := T^y(\varphi(x-y)).$$

The superscript y denotes the application of T to φ with respect to y. We have $T * \varphi \in C^\infty(\mathbb{R}^n)$ and $\partial^\alpha(T * \varphi) = (\partial^\alpha T) * \varphi = T * (\partial^\alpha \varphi)$.
Convolution of two distributions: For $T, S \in \mathcal{D}'(\mathbb{R}^n)$, one of which has compact support, we define $T * S \in \mathcal{D}'(\mathbb{R}^n)$ by

$$(T * S) * \varphi = T * (S * \varphi)$$

for all $\varphi \in C_0^\infty(\mathbb{R}^n)$.

Proposition A.10

1. *Let $T \in \mathcal{D}'(\mathbb{R}^n)$, $\varphi \in C_0^\infty(\mathbb{R}^n)$. Then we have $(T * \varphi)\check{} = \check{T} * \check{\varphi}$.*
2. *Let $T, S \in \mathcal{D}'(\mathbb{R}^n)$, one of which has compact support.*
   Then we have $(T * S)\check{} = \check{T} * \check{S}$.

PROOF: For 1. we have

$$(T * \varphi)\check{}(x) = (T * \varphi)(-x) = T^y(\varphi(-x-y))$$

$$(\check{T} * \check{\varphi})(x) = \check{T}^y(\check{\varphi}(x-y)) = \check{T}^y(\varphi(-x+y)) = T^y(\varphi(-x-y)).$$

To prove 2. we have for all $\varphi \in C_0^\infty(\mathbb{R}^n)$ by 1.

$$\begin{aligned}(T * S)\check{} * \check{\varphi} &= [(T * S) * \varphi]\check{} \\ &= [T * (S * \varphi)]\check{} \\ &= \check{T} * (S * \varphi)\check{} \\ &= \check{T} * (\check{S} * \check{\varphi}) \\ &= (\check{T} * \check{S}) * \check{\varphi},\end{aligned}$$

which proves 2. □

A.2 Fourier Transformation

We follow [69], Chapter VI., and [37], Chapters I. and II. There, one can also find the proofs of the results stated here. There are different conventions for the definition of the Fourier transform concerning the constant 2π. We use the definition of the Fourier transform given in [37], cf. the following definition.

Definition A.11 (Fourier transformation in L^1). *Let $f \in L^1(\mathbb{R}^n)$. We define the Fourier transform $\hat{f}$ by*

$$\hat{f}(\omega) := \int_{\mathbb{R}^n} f(x) e^{-i\omega^T x}\, dx.$$

If $\hat{f} \in L^1(\mathbb{R}^n)$, then the inversion formula $f(x) = (2\pi)^{-n} \int_{\mathbb{R}^n} \hat{f}(\omega) e^{ix^T\omega}\, d\omega$ holds.

L^1 is not mapped into itself under Fourier transformation. The Schwartz space of rapidly decreasing functions, however, will be mapped into itself by the Fourier transformation, cf. Proposition A.15.

Definition A.12 (The Schwartz space). *We define the following function space $\mathcal{S}(\mathbb{R}^n)$ of rapidly decreasing functions: $\varphi \in \mathcal{S}(\mathbb{R}^n)$ if and only if*

1. *$\varphi \in C^\infty(\mathbb{R}^n)$ and*
2. *for all multiindices α, β there is a constant $C_{\alpha,\beta}$ such that*

$$\sup_{x\in\mathbb{R}^n} \left| x^\beta D^\alpha \varphi(x) \right| \le C_{\alpha,\beta} \qquad \text{holds.}$$

Proposition A.13 (Properties of $\mathcal{S}(\mathbb{R}^n)$)

1. *$\mathcal{S}(\mathbb{R}^n) \subset L^p(\mathbb{R}^n)$ for all $p \ge 1$.*
2. *$C_0^\infty(\mathbb{R}^n) \subset \mathcal{S}(\mathbb{R}^n) \subset C^\infty(\mathbb{R}^n)$.*
3. *$C_0^\infty(\mathbb{R}^n)$ is dense in $\mathcal{S}(\mathbb{R}^n)$.*

Definition A.14 (Fourier transformation in $\mathcal{S}(\mathbb{R}^n)$).
Let $\varphi \in \mathcal{S}(\mathbb{R}^n)$. Then $\hat{\varphi}(\omega) = \int_{\mathbb{R}^n} \varphi(x) e^{-i\omega^T x}\, dx$.

Proposition A.15 *Let $\varphi \in \mathcal{S}(\mathbb{R}^n)$. Then*

$$\begin{aligned} \hat{\varphi} &\in \mathcal{S}(\mathbb{R}^n) \\ \varphi(x) &= (2\pi)^{-n} \int_{\mathbb{R}^n} \hat{\varphi}(\omega) e^{i\omega^T x}\, d\omega \text{ and} \\ \hat{\hat{\varphi}} &= (2\pi)^n \check{\varphi}. \end{aligned}$$

Moreover, $\check{\hat{\varphi}} = \hat{\check{\varphi}}$.

Proof: We show that last equation:

$$\check{\hat{\varphi}}(\omega) = \hat{\varphi}(-\omega) = \int_{\mathbb{R}^n} \varphi(x) e^{i\omega^T x}\, dx = \int_{\mathbb{R}^n} \varphi(-x) e^{-i\omega^T x}\, dx = \hat{\check{\varphi}}(\omega).$$

□

Proposition A.16 *Let $\varphi, \psi \in \mathcal{S}(\mathbb{R}^n)$. Then we have $\varphi * \psi \in \mathcal{S}(\mathbb{R}^n)$. Moreover, denoting $\langle \cdot, \cdot \rangle = \langle \cdot, \cdot \rangle_{L^2}$, we have*

$$\begin{aligned} \langle \hat{\varphi}, \psi \rangle &= \langle \varphi, \hat{\psi} \rangle \\ \langle \varphi, \overline{\psi} \rangle &= (2\pi)^{-n} \langle \hat{\varphi}, \overline{\hat{\psi}} \rangle \text{ (Parseval's formula)} \\ \widehat{\varphi * \psi} &= \hat{\varphi} \cdot \hat{\psi} \\ \widehat{\varphi \cdot \psi} &= (2\pi)^{-n} \hat{\varphi} * \hat{\psi}. \end{aligned}$$

Definition A.17 ($\mathcal{S}'(\mathbb{R}^n)$, Fourier transformation in $\mathcal{S}'(\mathbb{R}^n)$). *We define $\mathcal{S}'(\mathbb{R}^n)$ as the space of continuous linear operators on $\mathcal{S}(\mathbb{R}^n)$. Then*

$$\mathcal{E}'(\mathbb{R}^n) \subset \mathcal{S}'(\mathbb{R}^n) \subset \mathcal{D}'(\mathbb{R}^n).$$

Moreover, $C_0^\infty(\mathbb{R}^n)$ is dense in $\mathcal{S}'(\mathbb{R}^n)$.
We define the Fourier transformation for $T \in \mathcal{S}'(\mathbb{R}^n)$ by

$$\langle \hat{T}, \varphi \rangle := \langle T, \hat{\varphi} \rangle$$

for $\varphi \in \mathcal{S}(\mathbb{R}^n)$.

Proposition A.18 (Fourier transformation in $\mathcal{S}'(\mathbb{R}^n)$ – properties)

1. *For $T \in \mathcal{S}'(\mathbb{R}^n)$ we have $\hat{\hat{T}} = (2\pi)^n \check{T}$ and $\check{\hat{T}} = \hat{\check{T}}$.*
2. *For $T \in \mathcal{E}'(\mathbb{R}^n)$ we have $\hat{T}(\omega) = T^x(e^{-ix^T\omega})$; this is an analytic function with respect to ω which is polynomially bounded for $\omega \in \mathbb{R}^n$ (Theorem of Paley-Wiener).*
3. *For $T \in \mathcal{S}'(\mathbb{R}^n)$ we have $\overline{\hat{\check{T}}} = \hat{\overline{T}}$.*

4. For $T_1 \in \mathcal{S}'(\mathbb{R}^n)$ and $T_2 \in \mathcal{E}'(\mathbb{R}^n)$ we have $T_1 * T_2 \in \mathcal{S}'(\mathbb{R}^n)$ and

$$\widehat{T_1 * T_2} = \hat{T}_1 \cdot \hat{T}_2,$$
$$\widehat{\hat{T}_1 \cdot \hat{T}_2} = (2\pi)^n \check{T}_1 * \check{T}_2.$$

Note that the product $\hat{T}_1 \cdot \hat{T}_2$ is defined, since $\hat{T}_2 \in C^\infty(\mathbb{R}^n)$.

PROOF: We show 3.: For $\varphi \in \mathcal{S}(\mathbb{R}^n)$ we have $\overline{\hat{\varphi}(\omega)} = \int_{\mathbb{R}^n} e^{ix^T\omega}\overline{\varphi(x)}\, dx = \check{\hat{\overline{\varphi}}}(\omega)$. Hence,

$$\langle \overline{\hat{\check{T}}}, \varphi \rangle = \overline{\langle \hat{\check{T}}, \overline{\varphi} \rangle} = \overline{\langle T, \check{\hat{\overline{\varphi}}} \rangle} = \overline{\langle T, \overline{\hat{\varphi}} \rangle} = \langle \hat{\overline{T}}, \varphi \rangle.$$

4. The second formula follows from the first by Fourier transformation in $\mathcal{S}'(\mathbb{R}^n)$, 1. and Proposition A.10. □

Definition A.19 (Sobolev space). *We define for $s \in \mathbb{R}$ the* Sobolev space

$$H^s(\mathbb{R}^n) := \{u \in \mathcal{S}'(\mathbb{R}^n) \mid (1 + \|\omega\|^2)^{\frac{s}{2}} \hat{u}(\omega) \in L^2(\mathbb{R}^n)\}$$

equipped with the scalar product

$$\langle u, v \rangle := \int_{\mathbb{R}^n} (1 + \|\omega\|^2)^s \hat{u}(\omega) \overline{\hat{v}(\omega)}\, d\omega.$$

$H^s(\mathbb{R}^n)$ is a Hilbert space. For $s \in \mathbb{N}_0$, the Sobolev space $H^s(\mathbb{R}^n)$ coincides with the space

$$\{u \in L^2(\mathbb{R}^n) \mid D^\alpha u \in L^2(\mathbb{R}^n), |\alpha| \leq s\}$$

with scalar product $\langle u, v \rangle = \sum_{|\alpha| \leq s} \int_{\mathbb{R}^n} D^\alpha u(x) \overline{D^\alpha v(x)}\, dx$. The induced norms are equivalent to each other.

B

Data

B.1 Wendland Functions

In the following table we present the functions ψ_1 and ψ_2 for the Wendland functions $\psi_{3,1}(cr)$, $\psi_{4,2}(cr)$ and $\psi_{5,3}(cr)$. Note that these are the Wendland functions defined in Definition 3.9 up to a constant, cf. also Table 3.1 in Section 3.1.4. Note that $x_+ = x$ for $x \geq 0$ and $x_+ = 0$ for $x < 0$.

	$\psi_{3,1}(cr)$
$\psi(r)$	$(1-cr)_+^4[4cr+1]$
$\psi_1(r)$	$-20c^2(1-cr)_+^3$
$\psi_2(r)$	$60c^3\frac{1}{r}(1-cr)_+^2$

	$\psi_{4,2}(cr)$
$\psi(r)$	$(1-cr)_+^6[35(cr)^2+18cr+3]$
$\psi_1(r)$	$-56c^2(1-cr)_+^5[1+5cr]$
$\psi_2(r)$	$1680c^4(1-cr)_+^4$

	$\psi_{5,3}(cr)$
$\psi(r)$	$(1-cr)_+^8[32(cr)^3+25(cr)^2+8cr+1]$
$\psi_1(r)$	$-22c^2(1-cr)_+^7[16(cr)^2+7cr+1]$
$\psi_2(r)$	$528c^4(1-cr)_+^6[6cr+1]$

B.2 Figures

The parameters for the figures. The grid points are $x_0+\alpha\left(i+\frac{j}{2}, j\frac{\sqrt{3}}{2}\right)$ where $i,j\in\mathbb{Z}$ without the equilibrium ($i=j=0$) plus some additional points (add.), thus altogether N points. For the three-dimensional example we used the points $x_0+\alpha\left(i+\frac{j}{2}+\frac{l}{2}, j\frac{\sqrt{3}}{2}+\frac{l}{2\sqrt{3}}, l\sqrt{\frac{2}{3}}\right)$ with $i,j,l\in\mathbb{Z}$, excluding the equilibrium ($i=j=l=0$); α is proportional to the fill distance. This hexagonal grid and its generalization have been discussed at the beginning of Chapter 6.

The local Lyapunov function $\mathfrak{q}=\mathfrak{v}$ or $\mathfrak{q}=\mathfrak{d}$ was used, the sets $\{x \mid \mathfrak{q}(x)\le L\}$ and $\{x \mid q(x)\le R\}$ were calculated, where the Lyapunov function Q was approximated. Note that we did not add a suitable constant to q so that sometimes $R<0$. For interpolation of V via W we always used a Taylor polynomial of order $P=5$. The mixed interpolations in Figures 5.4 and 6.11 used a second grid X_M^0 with M points. The scaled Wendland function $\psi_{l,k}(cr)$ with $l=\lfloor\frac{n}{2}\rfloor+k+1$ was used as radial basis function.

Chemostat (1.1)

Fig.	k	c	α	add.	N	Q'	R	local	L
1.1 to 1.4	1	$\frac{1}{6}$	$\frac{1}{16}$	0	153	$-\|x\|^2$	-1.7	$\mathfrak{v}$	0.025

Example throughout the book (2.11)

Fig.	k	c	α	add.	N	M	Q'	R	local	L
2.1, 4.1, 4.2, 5.1l, 6.1r	2	$\frac{2}{3}$	0.4	2	24	–	$-\|x\|^2$	-0.95	$\mathfrak{v}$	0.09
4.3l, 4.4l, 5.1r	2	$\frac{2}{3}$	0.2	0	76	–	$-\|x\|^2$	-1.5	$\mathfrak{v}$	0.09
4.3r	2	$\frac{2}{3}$	0.2	0	76	–	$-\|x\|^2$	-1.5	$\mathfrak{v}$	0.045
4.4r	2	$\frac{2}{3}$	0.2	0	76	–	via W	0.2	–	–
5.2l	2	$\frac{2}{3}$	0.15	4	140	–	$-\|x\|^2$	-1.8	$\mathfrak{v}$	0.09
5.2r	2	$\frac{2}{3}$	0.1	0	312	–	$-\|x\|^2$	-3	$\mathfrak{v}$	0.09
5.3	2	$\frac{2}{3}$	0.075	0	484	–	$-\|x\|^2$	-3.4	$\mathfrak{v}$	0.09
5.4	2	$\frac{2}{3}$	0.2	0	70	10	$-\|x\|^2$	1.1	$\mathfrak{v}$	0.09

Speed-control (6.1)

Fig.	k	c	α	add.	N	Q'	R	local	L
6.2, 6.3m	–	–	–	–	–	–	–	$\mathfrak{d}$	0.035
6.3l, 6.4, 6.5	1	$\frac{5}{6}$	0.05	0	223	−1	−6	$\mathfrak{v}$	0.032

Toy example (6.2)

Fig.	k	c	α	add.	N	Q'	R	local	L
6.6r, 6.7l, 6.8l	1	$\frac{1}{2}$	0.3	4	126	$-\Vert x\Vert^2$	−2.9	$\mathfrak{v}$	0.35
6.7m	1	$\frac{1}{2}$	0.3	0	122	$-\Vert f(x)\Vert^2$	−0.65	$\mathfrak{v}$	0.35
6.7r	1	$\frac{1}{2}$	0.3	0	122	−1	−0.8	$\mathfrak{v}$	0.35
6.8m	1	$\frac{1}{2}$	0.3	0	122	via W	0.85	–	–

Three-dimensional Example (6.3)

Fig.	k	c	α	add.	N	Q'	R	local	L
6.9	1	$\frac{9}{20}$	0.35	0	137	$-\Vert x\Vert^2$	−0.595	$\mathfrak{v}$	0.05
6.10	1	$\frac{9}{20}$	0.35	0	137	via W	0.0654	–	–

van-der-Pol (6.4)

Fig.	k	c	α	add.	N	M	Q'	R	local	L
6.11	1	$\frac{10}{21}$	0.22	0	236	16	$-\Vert x\Vert^2$	1.6	$\mathfrak{v}$	1.15
6.12l	1	$\frac{10}{21}$	0.22	0	236	–	−2.5	−0.25	$\mathfrak{v}$	1.15

C

Notations

$'$	orbital derivative: $Q'(x) = \langle \nabla Q(x), f(x) \rangle$, cf. Definition 2.18
$\dot{}$	temporal derivative $\dot{x}(t) = \frac{d}{dt}x(t)$
$A(x_0)$	basin of attraction of x_0, cf. Definition 2.9
$\tilde{B}_r^Q(x_0)$	the set $\{x \in \mathbb{R}^n \mid Q(x) < r^2\}$ where Q is a Lyapunov function, cf. Definition 2.23
$\delta_{\tilde{x}}$	Dirac's delta-distribution at $\tilde{x} \in \mathbb{R}^n$, i.e. $\delta_{\tilde{x}} f(x) = f(\tilde{x})$, cf. Example A.4
$\mathfrak{d}(x)$	local Lyapunov function $\mathfrak{d}(x) = \|S(x - x_0)\|^2$ satisfying $\mathfrak{d}'(x) \leq 2(-\nu + \epsilon)\mathfrak{d}(x)$, cf. Lemma 2.28
$\mathcal{D}'(\mathbb{R}^n)$	space of distributions, cf. Definition A.2
$\mathcal{E}'(\mathbb{R}^n)$	space of distributions with compact support, cf. Definition A.5
f	$f \in C^\sigma(\mathbb{R}^n, \mathbb{R}^n)$, right-hand side of the ordinary differential equation $\dot{x} = f(x)$, cf. (2.1)
$\mathcal{F}$	native space of functions, cf. Definition 3.12
$\mathcal{F}^*$	dual of the native space, cf. Definition 3.12
$\mathfrak{h}(x)$	Taylor polynomial of $V(x)$ with $V'(x) = -\|x - x_0\|^2$, cf. Definition 2.52
$H^s(\mathbb{R}^n)$	Sobolev space, cf. Definition A.19
I	identity matrix $I = \text{diag}(1, 1, \ldots, 1)$
$\tilde{K}_r^Q(x_0)$	the set $\{x \in \mathbb{R}^n \mid Q(x) \leq r^2\}$ where Q is a Lyapunov function, cf. Definition 2.23
L	Lyapunov function with $L'(x) = -\bar{c}\, L(x)$ for $x \in A(x_0)$, cf. Corollary 2.40
$\mathfrak{n}(x)$	$\mathfrak{n}(x) = \mathfrak{h}(x) + M\|x - x_0\|^{2H}$, such that $\mathfrak{n}(x) > 0$ holds for $x \neq x_0$, cf. Definition 2.56
$\omega(x)$	ω-limit set of x, cf. Definition 2.12
Ω	non-characteristic hypersurface: $(n-1)$-dimensional manifold, often level set of a (local) Lyapunov function, cf. Definition 2.36
$\psi_{l,k}(r)$	Wendland function, cf. Definition 3.9

$\Psi(x)$	radial basis function, here $\Psi(x) = \psi_{l,k}(c\|x\|)$ with $c > 0$ and a Wendland function $\psi_{l,k}(r)$
$\mathcal{S}(\mathbb{R}^n)$	Schwartz space of rapidly decreasing functions, cf. Definition A.12
$\mathcal{S}'(\mathbb{R}^n)$	dual of the Schwartz space, cf. Definition A.17
$S_t\xi$	flow, solution $x(t)$ of $\dot{x} = f(x)$, $x(0) = \xi$, cf. Definition 2.1
T	Lyapunov function with $T'(x) = -\bar{c} < 0$ for $x \in A(x_0) \setminus \{x_0\}$ and $T(x) = H(x)$ for $x \in \Omega$, cf. Theorem 2.38
t	function which approximates T
V	Lyapunov function with $V'(x) = -p(x)$ for $x \in A(x_0)$, cf. Theorem 2.46
v	function which approximates V
$\mathfrak{v}$	local Lyapunov function satisfying $\langle \nabla \mathfrak{v}(x), Df(x_0)(x - x_0)\rangle = -\|x - x_0\|^2$, cf. Remark 2.34
V^*	Lyapunov function with $(V^*)'(x) = -p(x)$ for $x \in A(x_0) \setminus \{x_0\}$ and $V^*(x) = H(x)$ for $x \in \Omega$, cf. Proposition 2.51
x_0	equilibrium point of $\dot{x} = f(x)$, i.e. $f(x_0) = 0$, cf. Definition 2.6

References

1. R. Abraham & J. E. Marsden, *Foundations of Mechanics*, Addison-Wesley, Redwood City, CA, 1985 (2nd ed.).
2. H. Amann, *Ordinary Differential Equations. An Introduction to Nonlinear Analysis*, de Gruyter, Studies in Mathematics **13**, Berlin, 1990.
3. B. Aulbach, *Asymptotic Stability regions via Extensions of Zubov's Method – I*, Nonlinear Anal. **7** (1983) 12, 1431–1440.
4. B. Aulbach, *Asymptotic Stability regions via Extensions of Zubov's Method – II*, Nonlinear Anal. **7** (1983) 12, 1441–1454.
5. B. Aulbach, *Gewöhnliche Differenzialgleichungen*, Spektrum Akademischer Verlag, Heidelberg, 2004 (2nd ed.).
6. E. A. Barbašin & N. N. Krasovskiĭ, *On the existence of Liapunov functions in the case of asymptotic stability in the large*, Prikl. Mat. Mekh. **18** (1954), 345–350.
7. A. Barreiro, J. Aracil & D. Pagano, *Detection of attraction domains of non-linear systems using bifurcation analysis and Lyapunov functions*, Internat. J. Control **75** (2002), 314–327.
8. N. Bhatia, *On asymptotic stability in dynamical systems*, Math. Systems Theory **1** (1967), 113–128.
9. N. Bhatia & G. Szegö, *Stability Theory of Dynamical Systems*, Grundlehren der mathematischen Wissenschaften **161**, Springer, Berlin, 1970.
10. M. D. Buhmann, *Radial basis functions*, Acta Numer. **9** (2000), 1–38.
11. M. D. Buhmann, *Radial basis functions: theory and implementations*, Cambridge University Press, Cambridge, 2003.
12. F. Camilli, L. Grüne & F. Wirth, *A regularization of Zubov's equation for robust domains of attraction*, in: Nonlinear Control in the Year 2000, A. Isidori et al. (eds.), Lecture Notes in Control and Information Sciences **258**, Springer, London, 2000, 277–290.
13. F. Camilli, L. Grüne & F. Wirth, *A generalization of Zubov's method to perturbed systems*, SIAM J. Control Optim. **40** (2001) 2, 496–515.
14. H.-D. Chiang, M. W. Hirsch & F. F. Wu, *Stability Regions of Nonlinear Autonomous Dynamical Systems*, IEEE Trans. Automat. Control **33** (1988) 1, 16–27.
15. F. H. Clarke, Y. S. Ledayev & R. J. Stern, *Asymptotic stability and smooth Lyapunov functions*, J. Differential Equations **149** (1998), 69–115.

16. F. Fallside & M. R. Patel, *Step-response behaviour of a speed-control system with a back-e.m.f. nonlinearity*, Proc. IEE (London) **112** (1965) 10, 1979–1984.
17. C. Franke & R. Schaback, *Convergence order estimates of meshless collocation methods using radial basis functions*, Adv. Comput. Math. **8** (1998), 381–399.
18. C. Franke & R. Schaback, *Solving Partial Differential Equations by Collocation using Radial Basis Functions*, Appl. Math. Comput. **93** (1998) 1, 73–82.
19. R. Genesio, M. Tartaglia & A. Vicino, *On the Estimation of Asymptotic Stability Regions: State of the Art and New Proposals*, IEEE Trans. Automat. Control **30** (1985), 747–755.
20. P. Giesl, *Necessary Conditions for a Limit Cycle and its Basin of Attraction*, Nonlinear Anal. **56** (2004) 5, 643–677.
21. P. Giesl, *Construction of global Lyapunov functions using radial basis functions*, Habilitation Thesis, TU München, 2004.
22. P. Giesl, *Approximation of domains of attraction and Lyapunov functions using radial basis functions*, in: Proceedings of the NOLCOS 2004 Conference in Stuttgart, Germany, Vol. 2, 2004, 865–870.
23. P. Giesl, *Stepwise calculation of the basin of attraction in dynamical systems using radial basis functions*, in: "Algorithms for Approximation", A. Iske & J. Levesley (eds.), Springer, Heidelberg, 2007, 113–122.
24. P. Giesl, *Construction of a global Lyapunov function using radial basis functions with a single operator.* Discrete Cont. Dyn. Sys. Ser. B **7** (2007) No. 1, 101–124.
25. P. Giesl, *On the determination of the basin of attraction of discrete dynamical systems*, to appear in J. Difference Equ. Appl.
26. P. Giesl, *On the determination of the basin of attraction of a periodic orbit in two-dimensional systems.* Preprint, Munich, 2006.
27. P. Giesl & H. Wendland, *Meshless Collocation: Error estimates with Application to Dynamical Systems* (submitted to SIAM J. Numer. Anal.)
28. P. Giesl & H. Wendland, *Approximating the basin of attraction of time-periodic ODEs by meshless collocation.* Preprint Göttingen/Munich, 2006.
29. P. Glendinning, *Stability, instability and chaos: an introduction to the theory of nonlinear differential equations*, Cambridge University Press, Cambridge, 1994.
30. L. Grüne, *An adaptive grid scheme for the discrete Hamilton-Jacobi-Bellman equation*, Numer. Math. **75** (1997), 319–337.
31. L. Grüne, E. D. Sontag & F. Wirth, *Asymptotic stability equals exponential stability, and ISS equals finite energy gain – if you twist your eyes*, Systems Control Lett. **38** (1999) 2, 127–134.
32. S. Hafstein, *A constructive converse Lyapunov Theorem on Exponential Stability*, Discrete Contin. Dyn. Syst. **10** (2004) 3, 657–678.
33. W. Hahn, *Eine Bemerkung zur zweiten Methode von Ljapunov*, Math. Nachr. **14** (1956), 349–354.
34. W. Hahn, *Theorie und Anwendung der direkten Methode von Ljapunov*, Ergebnisse der Mathematik und ihrer Grenzgebiete **22**, Springer, Berlin, 1959.
35. W. Hahn, *Stability of Motion*, Springer, New York, 1967.
36. Ph. Hartman, *Ordinary Differential Equations*, Wiley, New York, 1964.
37. L. Hörmander, *Linear Partial Differential Operators*, Springer, Berlin, 1963.

38. A. Iske, *Reconstruction of Functions from Generalized Hermite-Birkhoff Data*, in: Approximation Theory VIII, Vol. 1: Approximation and Interpolation, Ch. Chui & L. Schumaker (eds.), 1995, 257–264.
39. A. Iske, *Perfect Centre Placement for Radial Basis Function Methods*, Technical Report, TUM-M9809, TU München, 1998.
40. A. Iske, *Scattered Data Modelling Using Radial Basis Functions*, in: Tutorials on Multiresolution in Geometric Modelling, A. Iske, E. Quak & M. Floater (eds.), Springer, 2002, 205–242.
41. L. B. Jocić, *Planar regions of attraction*, IEEE Trans. Automat. Control **27** (1982) 3, 708–710.
42. P. Julián, J. Guivant & A. Desages, *A parametrization of piecewise linear Lyapunov functions via linear programming*, Int. J. Control **72** (1999), 702–715.
43. E. J. Kansa, *Multiquadrics – a scattered data approximation scheme with applications to computational fluid-dynamics. i: Surface approximations and partial derivative estimates*, Comput. Math. Appl. **19** (1990), 127–145.
44. E. J. Kansa, *Multiquadrics – a scattered data approximation scheme with applications to computational fluid-dynamics. ii: Solutions to parabolic, hyperbolic and elliptic partial differential equations*, Comput. Math. Appl. **19** (1990), 147–161.
45. N. E. Kirin, R. A. Nelepin & V. N. Baidaev, *Construction of the Attraction Region by Zubov's method*, Differ. Equ. **17** (1981), 871–880.
46. N. N. Krasovskiĭ, *Stability of Motion*, Stanford University Press, Stanford, 1963. Translation of the russian original, Moscow, 1959.
47. Y. Lin, E. D. Sontag & Y. Wang, *A smooth converse Lyapunov theorem for robust stability*, SIAM J. Control Optim. **34** (1996), 124–160.
48. A. M. Lyapunov, *Problème général de la stabilité du mouvement*, Ann. Fac. Sci. Toulouse **9** (1907), 203–474. Translation of the russian version, published 1893 in Comm. Soc. math. Kharkow. Newly printed: Ann. of math. Stud. **17**, Princeton, 1949.
49. S. Marinossón, *Stability Analysis of Nonlinear Systems with Linear Programming. A Lyapunov Functions Based Approach*, Ph.D. thesis, University of Duisburg, Germany, 2002.
50. J. L. Massera, *On Liapounoff's Conditions of Stability*, Ann. of Math. **50** (1949) 2, 705–721.
51. F. J. Narcowich, J. D. Ward & H. Wendland, *Sobolev bounds on functions with scattered zeros, with applications to radial basis function surface fitting*, Math. Comp. **74** (2005), 643–763.
52. M. J. D. Powell, *The Theory of Radial Basis Function Approximation in 1990*, in: Advances in Numerical Analysis II: Wavelets, Subdivision Algorithms, and Radial Basis Functions, W. A. Light (ed.), Oxford University Press, 1992, 105–210.
53. R. Schaback, *Error estimates and condition numbers for radial basis function interpolation*, Adv. Comput. Math. **3** (1995), 251–264.
54. R. Schaback & H. Wendland, *Using compactly supported radial basis functions to solve partial differential equations*, in: Boundary Element Technology XIII (invited lecture), C. Chen, C. Brebbia & D. Pepper (eds.), WitPress, Southampton, Boston, 1999, 311–324.
55. H. R. Schwarz, *Numerische Mathematik*, Teubner, Stuttgart, 1988.
56. H. L. Smith & P. Waltman, *The theory of the chemostat*, Cambridge University Press, 1995.

57. E. D. SONTAG, *Mathematical control theory*, Springer, 1998 (2nd ed.).
58. A. R. TEEL & L. PRALY, *A smooth Lyapunov function from a class-KL estimate involving two positive semidefinite functions*, ESAIM Control Optim. Calc. Var. **5** (2000), 313–367.
59. A. VANNELLI & M. VIDYASAGAR, *Maximal Lyapunov Functions and Domains of Attraction for Autonomous Nonlinear Systems*, Automatica J. IFAC **21** (1985) 1, 69–80.
60. F. VERHULST, *Nonlinear Differential Equations and Dynamical Systems*, Springer, 1996.
61. H. WENDLAND, *Konstruktion und Untersuchung radialer Basisfunktionen mit kompaktem Träger*, Ph.D. thesis, University of Göttingen, Germany, 1996.
62. H. WENDLAND, *Error estimates for interpolation by compactly supported radial basis functions of minimal degree*, J. Approx. Theory **93** (1998), 258–272.
63. H. WENDLAND, *Scattered Data Approximation*, Cambridge Monographs on Applied and Computational Mathematics, Cambridge University Press, Cambridge, 2004.
64. F. WESLEY WILSON, JR., *The structure of the Level Surfaces of a Lyapunov Function*, J. Differential Equations **3** (1967), 323–329.
65. F. WESLEY WILSON, JR., *Smoothing Derivatives of Functions and Applications*, Trans. Amer. Math. Soc. **139** (1969), 413–428.
66. S. WIGGINS, *Introduction to Applied Nonlinear Dynamical Systems and Chaos*, Springer, 1990.
67. Z. WU, *Hermite-Birkhoff interpolation of scattered data by radial basis functions*, Approx. Theory Appl. **8** (1995), 283–292.
68. Z. WU & R. SCHABACK, *Local error estimates for radial basis function interpolation of scattered data*, IMA J. Numer. Anal. **13** (1993), 13–27.
69. K. YOSIDA, *Functional Analysis*, Springer, Berlin, 1974 (4th ed.).
70. V. I. ZUBOV, *Methods of A.M. Lyapunov and their Application*, P. Noordhoff, Groningen, 1964.

Index

Lecture Notes in Mathematics

For information about earlier volumes please contact your bookseller or Springer LNM Online archive: springerlink.com

Vol. 1711: W. Ricker, Operator Algebras Generated by Commuting Projećtions: A Vector Measure Approach (1999)
Vol. 1712: N. Schwartz, J. J. Madden, Semi-algebraic Function Rings and Reflectors of Partially Ordered Rings (1999)
Vol. 1713: F. Bethuel, G. Huisken, S. Müller, K. Steffen, Calculus of Variations and Geometric Evolution Problems. Cetraro, 1996. Editors: S. Hildebrandt, M. Struwe (1999)
Vol. 1714: O. Diekmann, R. Durrett, K. P. Hadeler, P. K. Maini, H. L. Smith, Mathematics Inspired by Biology. Martina Franca, 1997. Editors: V. Capasso, O. Diekmann (1999)
Vol. 1715: N. V. Krylov, M. Röckner, J. Zabczyk, Stochastic PDE's and Kolmogorov Equations in Infinite Dimensions. Cetraro, 1998. Editor: G. Da Prato (1999)
Vol. 1716: J. Coates, R. Greenberg, K. A. Ribet, K. Rubin, Arithmetic Theory of Elliptic Curves. Cetraro, 1997. Editor: C. Viola (1999)
Vol. 1717: J. Bertoin, F. Martinelli, Y. Peres, Lectures on Probability Theory and Statistics. Saint-Flour, 1997. Editor: P. Bernard (1999)
Vol. 1718: A. Eberle, Uniqueness and Non-Uniqueness of Semigroups Generated by Singular Diffusion Operators (1999)
Vol. 1719: K. R. Meyer, Periodic Solutions of the N-Body Problem (1999)
Vol. 1720: D. Elworthy, Y. Le Jan, X-M. Li, On the Geometry of Diffusion Operators and Stochastic Flows (1999)
Vol. 1721: A. Iarrobino, V. Kanev, Power Sums, Gorenstein Algebras, and Determinantal Loci (1999)
Vol. 1722: R. McCutcheon, Elemental Methods in Ergodic Ramsey Theory (1999)
Vol. 1723: J. P. Croisille, C. Lebeau, Diffraction by an Immersed Elastic Wedge (1999)
Vol. 1724: V. N. Kolokoltsov, Semiclassical Analysis for Diffusions and Stochastic Processes (2000)
Vol. 1725: D. A. Wolf-Gladrow, Lattice-Gas Cellular Automata and Lattice Boltzmann Models (2000)
Vol. 1726: V. Marić, Regular Variation and Differential Equations (2000)
Vol. 1727: P. Kravanja M. Van Barel, Computing the Zeros of Analytic Functions (2000)
Vol. 1728: K. Gatermann Computer Algebra Methods for Equivariant Dynamical Systems (2000)
Vol. 1729: J. Azéma, M. Émery, M. Ledoux, M. Yor (Eds.) Séminaire de Probabilités XXXIV (2000)
Vol. 1730: S. Graf, H. Luschgy, Foundations of Quantization for Probability Distributions (2000)
Vol. 1731: T. Hsu, Quilts: Central Extensions, Braid Actions, and Finite Groups (2000)
Vol. 1732: K. Keller, Invariant Factors, Julia Equivalences and the (Abstract) Mandelbrot Set (2000)
Vol. 1733: K. Ritter, Average-Case Analysis of Numerical Problems (2000)
Vol. 1734: M. Espedal, A. Fasano, A. Mikelić, Filtration in Porous Media and Industrial Applications. Cetraro 1998. Editor: A. Fasano. 2000.
Vol. 1735: D. Yafaev, Scattering Theory: Some Old and New Problems (2000)
Vol. 1736: B. O. Turesson, Nonlinear Potential Theory and Weighted Sobolev Spaces (2000)
Vol. 1737: S. Wakabayashi, Classical Microlocal Analysis in the Space of Hyperfunctions (2000)
Vol. 1738: M. Émery, A. Nemirovski, D. Voiculescu, Lectures on Probability Theory and Statistics (2000)
Vol. 1739: R. Burkard, P. Deuflhard, A. Jameson, J.-L. Lions, G. Strang, Computational Mathematics Driven by Industrial Problems. Martina Franca, 1999. Editors: V. Capasso, H. Engl, J. Periaux (2000)
Vol. 1740: B. Kawohl, O. Pironneau, L. Tartar, J.-P. Zolesio, Optimal Shape Design. Tróia, Portugal 1999. Editors: A. Cellina, A. Ornelas (2000)
Vol. 1741: E. Lombardi, Oscillatory Integrals and Phenomena Beyond all Algebraic Orders (2000)
Vol. 1742: A. Unterberger, Quantization and Non-holomorphic Modular Forms (2000)
Vol. 1743: L. Habermann, Riemannian Metrics of Constant Mass and Moduli Spaces of Conformal Structures (2000)
Vol. 1744: M. Kunze, Non-Smooth Dynamical Systems (2000)
Vol. 1745: V. D. Milman, G. Schechtman (Eds.), Geometric Aspects of Functional Analysis. Israel Seminar 1999-2000 (2000)
Vol. 1746: A. Degtyarev, I. Itenberg, V. Kharlamov, Real Enriques Surfaces (2000)
Vol. 1747: L. W. Christensen, Gorenstein Dimensions (2000)
Vol. 1748: M. Ruzicka, Electrorheological Fluids: Modeling and Mathematical Theory (2001)
Vol. 1749: M. Fuchs, G. Seregin, Variational Methods for Problems from Plasticity Theory and for Generalized Newtonian Fluids (2001)
Vol. 1750: B. Conrad, Grothendieck Duality and Base Change (2001)
Vol. 1751: N. J. Cutland, Loeb Measures in Practice: Recent Advances (2001)
Vol. 1752: Y. V. Nesterenko, P. Philippon, Introduction to Algebraic Independence Theory (2001)
Vol. 1753: A. I. Bobenko, U. Eitner, Painlevé Equations in the Differential Geometry of Surfaces (2001)
Vol. 1754: W. Bertram, The Geometry of Jordan and Lie Structures (2001)
Vol. 1755: J. Azéma, M. Émery, M. Ledoux, M. Yor (Eds.), Séminaire de Probabilités XXXV (2001)
Vol. 1756: P. E. Zhidkov, Korteweg de Vries and Nonlinear Schrödinger Equations: Qualitative Theory (2001)

Vol. 1757: R. R. Phelps, Lectures on Choquet's Theorem (2001)
Vol. 1758: N. Monod, Continuous Bounded Cohomology of Locally Compact Groups (2001)
Vol. 1759: Y. Abe, K. Kopfermann, Toroidal Groups (2001)
Vol. 1760: D. Filipović, Consistency Problems for Heath-Jarrow-Morton Interest Rate Models (2001)
Vol. 1761: C. Adelmann, The Decomposition of Primes in Torsion Point Fields (2001)
Vol. 1762: S. Cerrai, Second Order PDE's in Finite and Infinite Dimension (2001)
Vol. 1763: J.-L. Loday, A. Frabetti, F. Chapoton, F. Goichot, Dialgebras and Related Operads (2001)
Vol. 1764: A. Cannas da Silva, Lectures on Symplectic Geometry (2001)
Vol. 1765: T. Kerler, V. V. Lyubashenko, Non-Semisimple Topological Quantum Field Theories for 3-Manifolds with Corners (2001)
Vol. 1766: H. Hennion, L. Hervé, Limit Theorems for Markov Chains and Stochastic Properties of Dynamical Systems by Quasi-Compactness (2001)
Vol. 1767: J. Xiao, Holomorphic Q Classes (2001)
Vol. 1768: M. J. Pflaum, Analytic and Geometric Study of Stratified Spaces (2001)
Vol. 1769: M. Alberich-Carramiñana, Geometry of the Plane Cremona Maps (2002)
Vol. 1770: H. Gluesing-Luerssen, Linear Delay-Differential Systems with Commensurate Delays: An Algebraic Approach (2002)
Vol. 1771: M. Émery, M. Yor (Eds.), Séminaire de Probabilités 1967-1980. A Selection in Martingale Theory (2002)
Vol. 1772: F. Burstall, D. Ferus, K. Leschke, F. Pedit, U. Pinkall, Conformal Geometry of Surfaces in S^4 (2002)
Vol. 1773: Z. Arad, M. Muzychuk, Standard Integral Table Algebras Generated by a Non-real Element of Small Degree (2002)
Vol. 1774: V. Runde, Lectures on Amenability (2002)
Vol. 1775: W. H. Meeks, A. Ros, H. Rosenberg, The Global Theory of Minimal Surfaces in Flat Spaces. Martina Franca 1999. Editor: G. P. Pirola (2002)
Vol. 1776: K. Behrend, C. Gomez, V. Tarasov, G. Tian, Quantum Comohology. Cetraro 1997. Editors: P. de Bartolomeis, B. Dubrovin, C. Reina (2002)
Vol. 1777: E. García-Río, D. N. Kupeli, R. Vázquez-Lorenzo, Osserman Manifolds in Semi-Riemannian Geometry (2002)
Vol. 1778: H. Kiechle, Theory of K-Loops (2002)
Vol. 1779: I. Chueshov, Monotone Random Systems (2002)
Vol. 1780: J. H. Bruinier, Borcherds Products on O(2,1) and Chern Classes of Heegner Divisors (2002)
Vol. 1781: E. Bolthausen, E. Perkins, A. van der Vaart, Lectures on Probability Theory and Statistics. Ecole d' Eté de Probabilités de Saint-Flour XXIX-1999. Editor: P. Bernard (2002)
Vol. 1782: C.-H. Chu, A. T.-M. Lau, Harmonic Functions on Groups and Fourier Algebras (2002)
Vol. 1783: L. Grüne, Asymptotic Behavior of Dynamical and Control Systems under Perturbation and Discretization (2002)
Vol. 1784: L. H. Eliasson, S. B. Kuksin, S. Marmi, J.-C. Yoccoz, Dynamical Systems and Small Divisors. Cetraro, Italy 1998. Editors: S. Marmi, J.-C. Yoccoz (2002)
Vol. 1785: J. Arias de Reyna, Pointwise Convergence of Fourier Series (2002)
Vol. 1786: S. D. Cutkosky, Monomialization of Morphisms from 3-Folds to Surfaces (2002)
Vol. 1787: S. Caenepeel, G. Militaru, S. Zhu, Frobenius and Separable Functors for Generalized Module Categories and Nonlinear Equations (2002)
Vol. 1788: A. Vasil'ev, Moduli of Families of Curves for Conformal and Quasiconformal Mappings (2002)
Vol. 1789: Y. Sommerhäuser, Yetter-Drinfel'd Hopf algebras over groups of prime order (2002)
Vol. 1790: X. Zhan, Matrix Inequalities (2002)
Vol. 1791: M. Knebusch, D. Zhang, Manis Valuations and Prüfer Extensions I: A new Chapter in Commutative Algebra (2002)
Vol. 1792: D. D. Ang, R. Gorenflo, V. K. Le, D. D. Trong, Moment Theory and Some Inverse Problems in Potential Theory and Heat Conduction (2002)
Vol. 1793: J. Cortés Monforte, Geometric, Control and Numerical Aspects of Nonholonomic Systems (2002)
Vol. 1794: N. Pytheas Fogg, Substitution in Dynamics, Arithmetics and Combinatorics. Editors: V. Berthé, S. Ferenczi, C. Mauduit, A. Siegel (2002)
Vol. 1795: H. Li, Filtered-Graded Transfer in Using Non-commutative Gröbner Bases (2002)
Vol. 1796: J.M. Melenk, hp-Finite Element Methods for Singular Perturbations (2002)
Vol. 1797: B. Schmidt, Characters and Cyclotomic Fields in Finite Geometry (2002)
Vol. 1798: W.M. Oliva, Geometric Mechanics (2002)
Vol. 1799: H. Pajot, Analytic Capacity, Rectifiability, Menger Curvature and the Cauchy Integral (2002)
Vol. 1800: O. Gabber, L. Ramero, Almost Ring Theory (2003)
Vol. 1801: J. Azéma, M. Émery, M. Ledoux, M. Yor (Eds.), Séminaire de Probabilités XXXVI (2003)
Vol. 1802: V. Capasso, E. Merzbach, B. G. Ivanoff, M. Dozzi, R. Dalang, T. Mountford, Topics in Spatial Stochastic Processes. Martina Franca, Italy 2001. Editor: E. Merzbach (2003)
Vol. 1803: G. Dolzmann, Variational Methods for Crystalline Microstructure – Analysis and Computation (2003)
Vol. 1804: I. Cherednik, Ya. Markov, R. Howe, G. Lusztig, Iwahori-Hecke Algebras and their Representation Theory. Martina Franca, Italy 1999. Editors: V. Baldoni, D. Barbasch (2003)
Vol. 1805: F. Cao, Geometric Curve Evolution and Image Processing (2003)
Vol. 1806: H. Broer, I. Hoveijn. G. Lunther, G. Vegter, Bifurcations in Hamiltonian Systems. Computing Singularities by Gröbner Bases (2003)
Vol. 1807: V. D. Milman, G. Schechtman (Eds.), Geometric Aspects of Functional Analysis. Israel Seminar 2000-2002 (2003)
Vol. 1808: W. Schindler, Measures with Symmetry Properties (2003)
Vol. 1809: O. Steinbach, Stability Estimates for Hybrid Coupled Domain Decomposition Methods (2003)
Vol. 1810: J. Wengenroth, Derived Functors in Functional Analysis (2003)
Vol. 1811: J. Stevens, Deformations of Singularities (2003)
Vol. 1812: L. Ambrosio, K. Deckelnick, G. Dziuk, M. Mimura, V. A. Solonnikov, H. M. Soner, Mathematical Aspects of Evolving Interfaces. Madeira, Funchal, Portugal 2000. Editors: P. Colli, J. F. Rodrigues (2003)
Vol. 1813: L. Ambrosio, L. A. Caffarelli, Y. Brenier, G. Buttazzo, C. Villani, Optimal Transportation and its

Applications. Martina Franca, Italy 2001. Editors: L. A. Caffarelli, S. Salsa (2003)
Vol. 1814: P. Bank, F. Baudoin, H. Föllmer, L.C.G. Rogers, M. Soner, N. Touzi, Paris-Princeton Lectures on Mathematical Finance 2002 (2003)
Vol. 1815: A. M. Vershik (Ed.), Asymptotic Combinatorics with Applications to Mathematical Physics. St. Petersburg, Russia 2001 (2003)
Vol. 1816: S. Albeverio, W. Schachermayer, M. Talagrand, Lectures on Probability Theory and Statistics. Ecole d'Eté de Probabilités de Saint-Flour XXX-2000. Editor: P. Bernard (2003)
Vol. 1817: E. Koelink, W. Van Assche (Eds.), Orthogonal Polynomials and Special Functions. Leuven 2002 (2003)
Vol. 1818: M. Bildhauer, Convex Variational Problems with Linear, nearly Linear and/or Anisotropic Growth Conditions (2003)
Vol. 1819: D. Masser, Yu. V. Nesterenko, H. P. Schlickewei, W. M. Schmidt, M. Waldschmidt, Diophantine Approximation. Cetraro, Italy 2000. Editors: F. Amoroso, U. Zannier (2003)
Vol. 1820: F. Hiai, H. Kosaki, Means of Hilbert Space Operators (2003)
Vol. 1821: S. Teufel, Adiabatic Perturbation Theory in Quantum Dynamics (2003)
Vol. 1822: S.-N. Chow, R. Conti, R. Johnson, J. Mallet-Paret, R. Nussbaum, Dynamical Systems. Cetraro, Italy 2000. Editors: J. W. Macki, P. Zecca (2003)
Vol. 1823: A. M. Anile, W. Allegretto, C. Ringhofer, Mathematical Problems in Semiconductor Physics. Cetraro, Italy 1998. Editor: A. M. Anile (2003)
Vol. 1824: J. A. Navarro González, J. B. Sancho de Salas, $\mathscr{C}^\infty$ – Differentiable Spaces (2003)
Vol. 1825: J. H. Bramble, A. Cohen, W. Dahmen, Multiscale Problems and Methods in Numerical Simulations, Martina Franca, Italy 2001. Editor: C. Canuto (2003)
Vol. 1826: K. Dohmen, Improved Bonferroni Inequalities via Abstract Tubes. Inequalities and Identities of Inclusion-Exclusion Type. VIII, 113 p, 2003.
Vol. 1827: K. M. Pilgrim, Combinations of Complex Dynamical Systems. IX, 118 p, 2003.
Vol. 1828: D. J. Green, Gröbner Bases and the Computation of Group Cohomology. XII, 138 p, 2003.
Vol. 1829: E. Altman, B. Gaujal, A. Hordijk, Discrete-Event Control of Stochastic Networks: Multimodularity and Regularity. XIV, 313 p, 2003.
Vol. 1830: M. I. Gil', Operator Functions and Localization of Spectra. XIV, 256 p, 2003.
Vol. 1831: A. Connes, J. Cuntz, E. Guentner, N. Higson, J. E. Kaminker, Noncommutative Geometry, Martina Franca, Italy 2002. Editors: S. Doplicher, L. Longo (2004)
Vol. 1832: J. Azéma, M. Émery, M. Ledoux, M. Yor (Eds.), Séminaire de Probabilités XXXVII (2003)
Vol. 1833: D.-Q. Jiang, M. Qian, M.-P. Qian, Mathematical Theory of Nonequilibrium Steady States. On the Frontier of Probability and Dynamical Systems. IX, 280 p, 2004.
Vol. 1834: Yo. Yomdin, G. Comte, Tame Geometry with Application in Smooth Analysis. VIII, 186 p, 2004.
Vol. 1835: O.T. Izhboldin, B. Kahn, N.A. Karpenko, A. Vishik, Geometric Methods in the Algebraic Theory of Quadratic Forms. Summer School, Lens, 2000. Editor: J.-P. Tignol (2004)
Vol. 1836: C. Năstăsescu, F. Van Oystaeyen, Methods of Graded Rings. XIII, 304 p, 2004.
Vol. 1837: S. Tavaré, O. Zeitouni, Lectures on Probability Theory and Statistics. Ecole d'Eté de Probabilités de Saint-Flour XXXI-2001. Editor: J. Picard (2004)
Vol. 1838: A.J. Ganesh, N.W. O'Connell, D.J. Wischik, Big Queues. XII, 254 p, 2004.
Vol. 1839: R. Gohm, Noncommutative Stationary Processes. VIII, 170 p, 2004.
Vol. 1840: B. Tsirelson, W. Werner, Lectures on Probability Theory and Statistics. Ecole d'Eté de Probabilités de Saint-Flour XXXII-2002. Editor: J. Picard (2004)
Vol. 1841: W. Reichel, Uniqueness Theorems for Variational Problems by the Method of Transformation Groups (2004)
Vol. 1842: T. Johnsen, A. L. Knutsen, K_3 Projective Models in Scrolls (2004)
Vol. 1843: B. Jefferies, Spectral Properties of Noncommuting Operators (2004)
Vol. 1844: K.F. Siburg, The Principle of Least Action in Geometry and Dynamics (2004)
Vol. 1845: Min Ho Lee, Mixed Automorphic Forms, Torus Bundles, and Jacobi Forms (2004)
Vol. 1846: H. Ammari, H. Kang, Reconstruction of Small Inhomogeneities from Boundary Measurements (2004)
Vol. 1847: T.R. Bielecki, T. Björk, M. Jeanblanc, M. Rutkowski, J.A. Scheinkman, W. Xiong, Paris-Princeton Lectures on Mathematical Finance 2003 (2004)
Vol. 1848: M. Abate, J. E. Fornaess, X. Huang, J. P. Rosay, A. Tumanov, Real Methods in Complex and CR Geometry, Martina Franca, Italy 2002. Editors: D. Zaitsev, G. Zampieri (2004)
Vol. 1849: Martin L. Brown, Heegner Modules and Elliptic Curves (2004)
Vol. 1850: V. D. Milman, G. Schechtman (Eds.), Geometric Aspects of Functional Analysis. Israel Seminar 2002-2003 (2004)
Vol. 1851: O. Catoni, Statistical Learning Theory and Stochastic Optimization (2004)
Vol. 1852: A.S. Kechris, B.D. Miller, Topics in Orbit Equivalence (2004)
Vol. 1853: Ch. Favre, M. Jonsson, The Valuative Tree (2004)
Vol. 1854: O. Saeki, Topology of Singular Fibers of Differential Maps (2004)
Vol. 1855: G. Da Prato, P.C. Kunstmann, I. Lasiecka, A. Lunardi, R. Schnaubelt, L. Weis, Functional Analytic Methods for Evolution Equations. Editors: M. Iannelli, R. Nagel, S. Piazzera (2004)
Vol. 1856: K. Back, T.R. Bielecki, C. Hipp, S. Peng, W. Schachermayer, Stochastic Methods in Finance, Bressanone/Brixen, Italy, 2003. Editors: M. Fritelli, W. Runggaldier (2004)
Vol. 1857: M. Émery, M. Ledoux, M. Yor (Eds.), Séminaire de Probabilités XXXVIII (2005)
Vol. 1858: A.S. Cherny, H.-J. Engelbert, Singular Stochastic Differential Equations (2005)
Vol. 1859: E. Letellier, Fourier Transforms of Invariant Functions on Finite Reductive Lie Algebras (2005)
Vol. 1860: A. Borisyuk, G.B. Ermentrout, A. Friedman, D. Terman, Tutorials in Mathematical Biosciences I. Mathematical Neurosciences (2005)
Vol. 1861: G. Benettin, J. Henrard, S. Kuksin, Hamiltonian Dynamics – Theory and Applications, Cetraro, Italy, 1999. Editor: A. Giorgilli (2005)
Vol. 1862: B. Helffer, F. Nier, Hypoelliptic Estimates and Spectral Theory for Fokker-Planck Operators and Witten Laplacians (2005)

Vol. 1863: H. Führ, Abstract Harmonic Analysis of Continuous Wavelet Transforms (2005)
Vol. 1864: K. Efstathiou, Metamorphoses of Hamiltonian Systems with Symmetries (2005)
Vol. 1865: D. Applebaum, B.V. R. Bhat, J. Kustermans, J. M. Lindsay, Quantum Independent Increment Processes I. From Classical Probability to Quantum Stochastic Calculus. Editors: M. Schürmann, U. Franz (2005)
Vol. 1866: O.E. Barndorff-Nielsen, U. Franz, R. Gohm, B. Kümmerer, S. Thorbjønsen, Quantum Independent Increment Processes II. Structure of Quantum Lévy Processes, Classical Probability, and Physics. Editors: M. Schürmann, U. Franz, (2005)
Vol. 1867: J. Sneyd (Ed.), Tutorials in Mathematical Biosciences II. Mathematical Modeling of Calcium Dynamics and Signal Transduction. (2005)
Vol. 1868: J. Jorgenson, S. Lang, $Pos_n(R)$ and Eisenstein Series. (2005)
Vol. 1869: A. Dembo, T. Funaki, Lectures on Probability Theory and Statistics. Ecole d'Eté de Probabilités de Saint-Flour XXXIII-2003. Editor: J. Picard (2005)
Vol. 1870: V.I. Gurariy, W. Lusky, Geometry of Müntz Spaces and Related Questions. (2005)
Vol. 1871: P. Constantin, G. Gallavotti, A.V. Kazhikhov, Y. Meyer, S. Ukai, Mathematical Foundation of Turbulent Viscous Flows, Martina Franca, Italy, 2003. Editors: M. Cannone, T. Miyakawa (2006)
Vol. 1872: A. Friedman (Ed.), Tutorials in Mathematical Biosciences III. Cell Cycle, Proliferation, and Cancer (2006)
Vol. 1873: R. Mansuy, M. Yor, Random Times and Enlargements of Filtrations in a Brownian Setting (2006)
Vol. 1874: M. Yor, M. Émery (Eds.), In Memoriam Paul-André Meyer - Séminaire de probabilités XXXIX (2006)
Vol. 1875: J. Pitman, Combinatorial Stochastic Processes. Ecole d'Eté de Probabilités de Saint-Flour XXXII-2002. Editor: J. Picard (2006)
Vol. 1876: H. Herrlich, Axiom of Choice (2006)
Vol. 1877: J. Steuding, Value Distributions of L-Functions (2007)
Vol. 1878: R. Cerf, The Wulff Crystal in Ising and Percolation Models, Ecole d'Eté de Probabilités de Saint-Flour XXXIV-2004. Editor: Jean Picard (2006)
Vol. 1879: G. Slade, The Lace Expansion and its Applications, Ecole d'Eté de Probabilités de Saint-Flour XXXIV-2004. Editor: Jean Picard (2006)
Vol. 1880: S. Attal, A. Joye, C.-A. Pillet, Open Quantum Systems I, The Hamiltonian Approach (2006)
Vol. 1881: S. Attal, A. Joye, C.-A. Pillet, Open Quantum Systems II, The Markovian Approach (2006)
Vol. 1882: S. Attal, A. Joye, C.-A. Pillet, Open Quantum Systems III, Recent Developments (2006)
Vol. 1883: W. Van Assche, F. Marcellàn (Eds.), Orthogonal Polynomials and Special Functions, Computation and Application (2006)
Vol. 1884: N. Hayashi, E.I. Kaikina, P.I. Naumkin, I.A. Shishmarev, Asymptotics for Dissipative Nonlinear Equations (2006)
Vol. 1885: A. Telcs, The Art of Random Walks (2006)
Vol. 1886: S. Takamura, Splitting Deformations of Degenerations of Complex Curves (2006)
Vol. 1887: K. Habermann, L. Habermann, Introduction to Symplectic Dirac Operators (2006)
Vol. 1888: J. van der Hoeven, Transseries and Real Differential Algebra (2006)
Vol. 1889: G. Osipenko, Dynamical Systems, Graphs, and Algorithms (2006)
Vol. 1890: M. Bunge, J. Funk, Singular Coverings of Toposes (2006)
Vol. 1891: J.B. Friedlander, D.R. Heath-Brown, H. Iwaniec, J. Kaczorowski, Analytic Number Theory, Cetraro, Italy, 2002. Editors: A. Perelli, C. Viola (2006)
Vol. 1892: A. Baddeley, I. Bárány, R. Schneider, W. Weil, Stochastic Geometry, Martina Franca, Italy, 2004. Editor: W. Weil (2007)
Vol. 1893: H. Hanßmann, Local and Semi-Local Bifurcations in Hamiltonian Dynamical Systems, Results and Examples (2007)
Vol. 1894: C.W. Groetsch, Stable Approximate Evaluation of Unbounded Operators (2007)
Vol. 1895: L. Molnár, Selected Preserver Problems on Algebraic Structures of Linear Operators and on Function Spaces (2007)
Vol. 1896: P. Massart, Concentration Inequalities and Model Selection, Ecole d'Eté de Probabilités de Saint-Flour XXXIII-2003. Editor: J. Picard (2007)
Vol. 1897: R. Doney, Fluctuation Theory for Lévy Processes, Ecole d'Eté de Probabilités de Saint-Flour-2005. Editor: J. Picard (2007)
Vol. 1898: H.R. Beyer, Beyond Partial Differential Equations, On linear and Quasi-Linear Abstract Hyperbolic Evolution Equations (2007)
Vol. 1899: Séminaire de Probabilités XL. Editors: C. Donati-Martin, M. Émery, A. Rouault, C. Stricker (2007)
Vol. 1900: E. Bolthausen, A. Bovier (Eds.), Spin Glasses (2007)
Vol. 1901: O. Wittenberg, Intersections de deux quadriques et pinceaux de courbes de genre 1, Intersections of Two Quadrics and Pencils of Curves of Genus 1 (2007)
Vol. 1902: A. Isaev, Lectures on the Automorphism Groups of Kobayashi-Hyperbolic Manifolds (2007)
Vol. 1903: G. Kresin, V. Maz'ya, Sharp Real-Part Theorems (2007)
Vol. 1904: P. Giesl, Construction of Global Lyapunov Functions Using Radial Basis Functions (2007)

Recent Reprints and New Editions

Vol. 1618: G. Pisier, Similarity Problems and Completely Bounded Maps. 1995 – 2nd exp. edition (2001)
Vol. 1629: J.D. Moore, Lectures on Seiberg-Witten Invariants. 1997 – 2nd edition (2001)
Vol. 1638: P. Vanhaecke, Integrable Systems in the realm of Algebraic Geometry. 1996 – 2nd edition (2001)
Vol. 1702: J. Ma, J. Yong, Forward-Backward Stochastic Differential Equations and their Applications. 1999 – Corr. 3rd printing (2005)
Vol. 830: J.A. Green, Polynomial Representations of GL_n, with an Appendix on Schensted Correspondence and Littelmann Paths by K. Erdmann, J.A. Green and M. Schocker 1980 – 2nd corr. and augmented edition (2007)

4. Careful preparation of the manuscripts will help keep production time short besides ensuring satisfactory appearance of the finished book in print and online. After acceptance of the manuscript authors will be asked to prepare the final LaTeX source files (and also the corresponding dvi-, pdf- or zipped ps-file) together with the final printout made from these files. The LaTeX source files are essential for producing the full-text online version of the book (see http://www.springerlink.com/openurl.asp?genre=journal&issn=0075-8434 for the existing online volumes of LNM).

   The actual production of a Lecture Notes volume takes approximately 8 weeks.

5. Authors receive a total of 50 free copies of their volume, but no royalties. They are entitled to a discount of 33.3 % on the price of Springer books purchased for their personal use, if ordering directly from Springer.

6. Commitment to publish is made by letter of intent rather than by signing a formal contract. Springer-Verlag secures the copyright for each volume. Authors are free to reuse material contained in their LNM volumes in later publications: A brief written (or e-mail) request for formal permission is sufficient.

Addresses:

Professor J.-M. Morel, CMLA,
École Normale Supérieure de Cachan,
61 Avenue du Président Wilson, 94235 Cachan Cedex, France
E-mail: Jean-Michel.Morel@cmla.ens-cachan.fr

Professor F. Takens, Mathematisch Instituut,
Rijksuniversiteit Groningen, Postbus 800,
9700 AV Groningen, The Netherlands
E-mail: F.Takens@math.rug.nl

Professor B. Teissier, Institut Mathématique de Jussieu,
UMR 7586 du CNRS, Équipe "Géométrie et Dynamique",
175 rue du Chevaleret
75013 Paris, France
E-mail: teissier@math.jussieu.fr

For the "Mathematical Biosciences Subseries" of LNM:

Professor P. K. Maini, Center for Mathematical Biology,
Mathematical Institute, 24-29 St Giles,
Oxford OX1 3LP, UK
E-mail : maini@maths.ox.ac.uk

Springer, Mathematics Editorial, Tiergartenstr. 17,
69121 Heidelberg, Germany,
Tel.: +49 (6221) 487-8410
Fax: +49 (6221) 487-8355
E-mail: lnm@springer-sbm.com